T0180061

Practical NMR
Relaxation
for Chemists

Practical NMR Relaxation for Chemists

Vladimir I. Bakhmutov

Texas A&M University

John Wiley & Sons, Ltd

Copyright © 2004   John Wiley & Sons Ltd, The Atrium, Southern Gate, Chichester,
West Sussex PO19 8SQ, England

Telephone (+44) 1243 779777

Email (for orders and customer service enquiries): cs-books@wiley.co.uk
Visit our Home Page on www.wileyeurope.com or www.wiley.com

All Rights Reserved. No part of this publication may be reproduced, stored in a retrieval system or transmitted in any form or by any means, electronic, mechanical, photocopying, recording, scanning or otherwise, except under the terms of the Copyright, Designs and Patents Act 1988 or under the terms of a licence issued by the Copyright Licensing Agency Ltd, 90 Tottenham Court Road, London W1T 4LP, UK, without the permission in writing of the Publisher. Requests to the Publisher should be addressed to the Permissions Department, John Wiley & Sons Ltd, The Atrium, Southern Gate, Chichester, West Sussex PO19 8SQ, England, or emailed to permreq@wiley.co.uk, or faxed to (+44) 1243 770620.

This publication is designed to provide accurate and authoritative information in regard to the subject matter covered. It is sold on the understanding that the Publisher is not engaged in rendering professional services. If professional advice or other expert assistance is required, the services of a competent professional should be sought.

Other Wiley Editorial Offices

John Wiley & Sons Inc., 111 River Street, Hoboken, NJ 07030, USA

Jossey-Bass, 989 Market Street, San Francisco, CA 94103-1741, USA

Wiley-VCH Verlag GmbH, Boschstr. 12, D-69469 Weinheim, Germany

John Wiley & Sons Australia Ltd, 33 Park Road, Milton, Queensland 4064, Australia

John Wiley & Sons (Asia) Pte Ltd, 2 Clementi Loop #02-01, Jin Xing Distripark, Singapore 129809

John Wiley & Sons Canada Ltd, 22 Worcester Road, Etobicoke, Ontario, Canada M9W 1L1

Wiley also publishes its books in a variety of electronic formats. Some content that appears in print may not be available in electronic books.

Library of Congress Cataloging-in-Publication Data

Bakhmutov, Vladimir I.–
    Practical NMR relaxation for chemists / Vladimir
I. Bakhmutov.
        p. cm.
    Includes bibliographical references and index.
    ISBN 0-470-09445-1 (cloth : alk. paper) – ISBN 0-470-09446-X (pbk. :
alk. paper)
    1. Nuclear magnetic resonance. 2. Relaxation (Nuclear physics) I.
Title.
    QD581.B35 2004
    538'.362'02454--dc22

                        2004014374

British Library Cataloguing in Publication Data

A catalogue record for this book is available from the British Library

ISBN 0-470-09445-1 (HB)
        0-470-09446-X (PB)

Typeset in 10/12pt Palatino by Laserwords Private Limited, Chennai, India

Contents

Preface

Nuclear magnetic resonance, discovered by Bloch and Purcell in 1946, is widely used as a powerful analytical method in different fields of modern science, medicine and industry. It is difficult to overestimate the role of NMR in fundamental and applied chemistry where practically each chemical study, from the simplest organic molecules to complex molecular systems such as proteins, leans upon the data obtained by NMR experiments, carried out on different nuclei. Moreover, modern NMR is an indispensable tool for the practicing chemist.

Historically, since the appearance of the first commercial NMR instruments, practical applications of NMR split up two different spheres: NMR spectroscopy and NMR relaxation. The first domain deals with NMR spectra which show the number of distinct nuclei in investigated samples. In other words, the spectra are directly connected with structures of compounds. This circumstance explains the popularity of NMR spectroscopy among chemists. The resulting data, collected by nuclear relaxation experiments are time dependent. For this reason, relaxation is related to the dynamics of investigated objects: rotational or translational motions in liquids and solids, phase transitions in the solid state, spin dynamics in the solid state and molecular mobility in liquid crystals. All these problems are within the ambit of molecular physics, the physics of solids and materials science.

Spin relaxation has attracted the attention of physicists from the early days of NMR, and the theory of the observed phenomena has been rapidly developed and applied, first of all in studies of solids. Nowadays, NMR relaxation plays an important role in biophysics where it helps to characterize motions in complex biological macromolecules such as proteins and nucleic acids. Thus, these studies throw light upon the biological activity of macromolecules and on the change of the activity upon binding with other molecules.

H. Günther and J. Kowalewski note that NMR relaxation studies are almost as old as the NMR method itself. Moreover spin relaxation has played a major role in traditional NMR spectroscopy. Actually, as we will see below, long T_1 and short T_2 times make NMR spectroscopic experiments very difficult, sometimes impossible. H. Günther reminds us that the first attempt to demonstrate the NMR phenomenon for ^1H and ^7Li nuclei, performed by Gorter in 1936, was prevented by relaxation. The popularity of relaxation experiments among chemists is still not very high. Partly this could be explained by the interests of synthetic chemists, working on the design of new molecular systems and their structures. However, relaxation experiments can provide unique structural information, particularly in solutions where other structural approaches are unavailable.

The present book is not a scientific monograph, and does not claim to be a complete account. Its task is simple: to show in practice how relaxation experiments on protons, deuterons or other nuclei can be applied for qualitative structural diagnostics in solutions, quantitative structural determinations, recognitions of weak intermolecular interactions and studies of molecular mobility. Focusing on methodical aspects and discussing the possible sources of errors in relaxation time determinations and their interpretations, we consciously avoid the complex quantum mechanical descriptions. We use macroscopic equations, which are converted into simple forms, convenient for applications. Thus, the reader does not need any special knowledge of physics and NMR. In addition the first chapters of the book give the theoretical basics of nuclear relaxation and explain how and why nuclei relax. Finally, we believe that the small size of the book and its simplicity will stimulate further learning about nuclear relaxation and wide applications of the NMR relaxation technique in chemistry.

Vladimir I. Bakhmutov
Department of Chemistry, Texas A&M University, College Station, USA

1 How and Why Nuclei Relax

In spite of the constant technical improvements to NMR spectrometers and developments in NMR experiments [1–4], their physical meaning may be defined as an excitation of nuclei, placed in an external magnetic field, by radiofrequency irradiation, followed by registration of absorbed energy as NMR signals. The signals form NMR spectra that are recorded as plots of the line intensity versus frequency. Dispositions of resonance lines in the spectra, characterized by chemical shifts (ppm), and their splitting due to spin–spin coupling, measured in Hz (Figure 1.1), depend on the electronic environments of nuclei. Integral intensities of the signals are proportional to the number of resonating nuclei. That is why NMR spectra are directly related to molecular structures.

We illustrate a scheme of the simplest NMR experiment, resulting in a so-called one-dimensional (1D) NMR spectrum. As seen from Figure 1.2, the experiment consists of several time sections: (i) an initial time delay RD (the so-called relaxation delay); (ii) a radiofrequency pulse (RFP) exciting nuclei; and (iii) collection of NMR data as free induction decays (FID) during the acquisition time (AT). Sections (i)–(iii) can be repeated in order to accumulate the NMR data if necessary. The collected NMR data are *time dependent* and therefore they can be expressed as a function of time $f(t)$. In contrast, NMR spectra represent the *frequency-dependent* data $F(v)$. Thus, NMR decays

Practical NMR Relaxation for Chemists Vladimir I. Bakhmutov
© 2004 John Wiley & Sons, Ltd   ISBNs: 0-470-09445-1 (HB); 0-470-09446-X (PB)

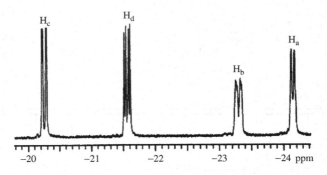

Figure 1.1 Hydride region of the ^{1}H NMR spectrum of a C_6D_6 solution of the complex $[Ir_2(\mu\text{-Ha})(\mu\text{-Pz})_2(Hb)(Hc)(Hd)(NCCH_3)(PPr^i_3)_2]$ containing one bridging and three terminal hydride ligands. (Reproduced with permission from E. Sola *et al.* *Orgamometallics* 1998; **17**: 693. © 1998 American Chemical Society)

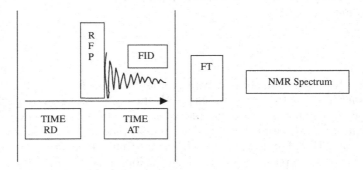

Figure 1.2 Schematic presentation of the simplest 1D NMR experiment where typical durations of RFP, RD and AT are of ∼5–10 µs, 1–4 s, and 1–3 s, respectively

require the mathematical procedure of Fourier transformation (FT) [2, 3]:

$$F(v) = \int f(t) \exp(-i2\pi vt) \, dt \qquad (1.1)$$

$$f(t) = \int F(v) \exp(+i2\pi vt) \, dv$$

converting the frequency domain into the time domain and vice versa.

Any two-dimensional (2D) NMR experiment [2, 3] adds a second frequency axis. One of the simplest 2D NMR experiments, named ^{1}H–^{1}H COSY, is shown in Figure 1.3. Here, after the action of the first pulse, a nuclear system develops by proton–proton spin–spin coupling during time t_1. Then, the first set of time-dependent data $f(t_2)$ is collected during time t_2 after the action of the second radiofrequency pulse. The second set of time-dependent data $f(t_1)$ can be collected if t_1 is variable. Finally, double Fourier transformation with respect to t_1 and t_2 creates two frequency domains. The resulting 2D NMR

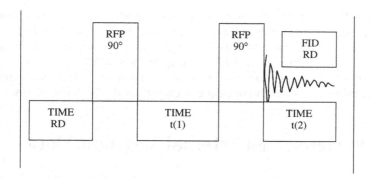

Figure 1.3 Pulse sequence for the 2D homonuclear proton–proton correlation (COSY) NMR experiment. The second pulse can be 90° or 45°

spectrum is a square plot where the diagonal shows the 1D ^{1}H NMR spectrum and cross peaks appear, due to scalar spin–spin coupling. Additionally, the coordinate 'intensity' in such a spectrum is located perpendicular to the plane formed by the frequency coordinates. The examples of ^{1}H 2D NMR spectra, ^{1}H–^{1}H COSY and ^{1}H–^{1}H NOESY, are shown in Figure 1.4.

In the context of this book, we are not concerned with traditional NMR spectra and their interpretations on the basis of numerous spectrum–structure relations. These aspects are well treated, for example, in the book by

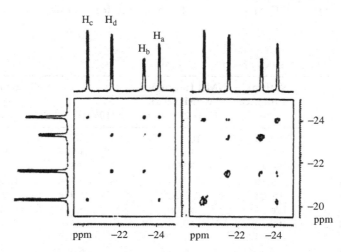

Figure 1.4 Hydride region of the ^{1}H COSY (left) and NOESY (right) NMR spectrum of a C_6D_6 solution of the complex [Ir$_2$(μ-Ha)(μ-Pz)$_2$(Hb)(Hc)(Hd)(NCCH$_3$)(PPr$^i_3$)$_2$] containing one bridging and three terminal hydride ligands. (Reproduced with permission from E. Sola *et al. Orgamometallics* 1998; **17**: 693. © 1998 American Chemical Society)

Harris [5]. We will be interested in time-dependent data, which are governed by nuclear relaxation. However it should be emphasized that, even in traditional NMR experiments, nuclear relaxation plays a major role. In fact, as it follows from Figures 1.2 and 1.3, the registration of NMR signals is impossible if relaxation times are infinitely large. To understand this statement better, we start from the theoretical basics of the NMR phenomenon.

1.1 Nucleus in an External Magnetic Field

Quantum mechanical formalism [1] is the best way to explain clearly the behavior of nuclei in a strong external magnetic field. According to this formalism, any nucleus is magnetically active if it posses a nonzero *angular moment P*, which is responsible for the appearance of the *nuclear magnetic moment* μ:

$$\mu = \gamma P \tag{1.2}$$

capable of interaction with the external magnetic field. Coefficient γ in Equation (1.2), the nuclear magnetogyric ratio, is one of the fundamental magnetic constants of nuclei, dependent on their nature (Table 1.1). As we show below, the γ values dictate frequencies for observation of NMR. Sensitivity in NMR experiments also depends on the γ values. Comparison of the magnetic properties and natural abundance of isotopes in Table 1.1 shows that ^{1}H and ^{19}F nuclei are most convenient for chemical investigations by the NMR method.

Table 1.1 NMR properties of some nuclei [5]

Nucleus	Spin I	Natural abundance (%)	NMR frequency, v_0 (MHz) at $B_0 = 2.3488$ T	γ (10^7 rad T^{-1} s^{-1})	Sensitivity relative to ^{1}H
^{1}H	1/2	99.98	100	26.752	1.000
^{2}H	1	0.016	15.35	4.107	1.45×10^{-6}
^{11}B	3/2	80.42	32.08	8.584	0.133
^{13}C	1/2	1.108	25.14	6.728	1.76×10^{-4}
^{14}N	1	99.63	7.22	1.934	1.00×10^{-3}
^{19}F	1/2	100	94.08	25.18	0.834
^{31}P	1/2	100	40.48	10.84	0.065
^{17}O	5/2	0.037	13.557	−3.628	1.08×10^{-5}
^{93}Nb	9/2	100	24.549	6.567	0.487
^{117}Sn	1/2	7.61	35.63	−9.578	3.49×10^{-3}
^{119}Sn	1/2	8.58	37.29	−10.02	4.51×10^{-3}
^{199}Hg	1/2	16.84	17.91	4.815	9.82×10^{-4}
^{205}Tl	1/2	70.5	57.63	15.589	0.140

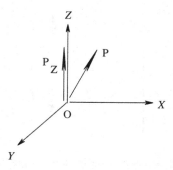

Figure 1.5 Projection of the nuclear angular moment P on the Z-axis. The external magnetic field is usually applied along this direction

The magnitudes of the P and the μ are *quantized*. In other words, projections of the angular nuclear moment P_Z on the Z-axis (Figure 1.5) can be written as:

$$P_Z = \hbar\, m_I \tag{1.3}$$

$$\hbar = h/2\pi$$

where h is Planck's constant ($h/2\pi = 1.05457266 \times 10^{-34}$ J s) and m_I is the magnetic quantum number. The value of m_I depends on the nuclear spin I, and takes values from I to $-I$:

$$I, I-1, I-2 \ldots -I \tag{1.4}$$

According to the quantum mechanical formalism, I is a multiple of $1/2$. Then, for example, at $I = 1/2$ the angular and magnetic moments can be expressed by:

$$P_Z = \pm(1/2)\hbar$$

$$\mu_Z = \pm(1/2)\gamma\hbar \tag{1.5}$$

Hence, the μ_Z (or P_Z) projections, imagined as the vectors, can take only two 'permitted' spatial orientations: *parallel* or *antiparallel* with respect to the Z-axis. In the absence of an external magnetic field B_0, both orientations are energetically equivalent and thus nuclei occupy a single energy level. The situation changes when a strong magnetic field is applied along the Z-axis and the orientations become nonequivalent. Thus the initial energy level undergoes so-called Zeeman splitting (Figure 1.6). The apparent energy difference ΔE is proportional to the μ_Z magnitude and the strength of the applied magnetic field B_0:

$$\Delta E = 2\,\mu_Z B_0 \tag{1.6}$$

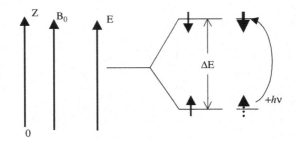

Figure 1.6 Zeeman energy levels in the external magnetic field B_0 for nuclear spin $I = 1/2$

According to the Bohr rules, a minimal energy difference is formulated as *one quantum*, $\Delta E = h\nu$, and then one can write:

$$h\nu_0 = 2\,\mu_Z B_0 = \gamma\hbar B_0$$
$$\nu_0 = \gamma B_0 \tag{1.7}$$

These fundamental equations formulate the resonance conditions: a nucleus with the magnetogyric ratio γ and spin of $1/2$, placed into an external magnetic field B_0, undergoes a single-quantum transition (the m_I number changes from $-1/2$ to $+1/2$) between the Zeeman energy levels from a low energy to a high energy when it is irradiated with frequency ν_0. As can be seen, the Larmor frequency ν_0, is dependent on the nature of the nuclei (see the γ) and the strength of the applied magnetic field B_0. It is obvious that the above nuclear transition is accompanied by an energy absorption ΔE, registered as an NMR signal.

A macroscopic sample, placed in the external magnetic field, represents an ensemble of nuclei. These nuclei populate the Zeeman energy levels proportionally to the factor:

$$\exp(-\Delta E/kT) \tag{1.8}$$

where T is the temperature and k is the Boltzmann constant (1.380658×10^{-23} J/K). For this reason, the level with a lower energy in Figure 1.7 will be more populated, thus creating the conditions for observation of NMR.

To show the behavior of nuclei with spin numbers larger than $1/2$, consider the Zeeman splitting for nuclei ^{2}H or ^{14}N ($I = 1$). According to the formalism, these nuclei, placed in the magnetic field, give three energy levels, corresponding to the μ_Z values $+\gamma\hbar(1)$, 0 and $-\gamma\hbar(1)$. It is obvious that the levels are energetically equidistant. In addition, the double-quantum nuclear transitions (when the m_I value changes from $+1$ to -1) are forbidden. Thus, in spite of the presence of three energy levels, identical nuclei give a single resonance line in the NMR spectrum.

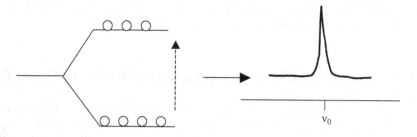

Figure 1.7 Difference in populations of the Zeeman energy levels leading to observation of a NMR signal at irradiation of a sample

1.2 Spin–Lattice and Spin–Spin Nuclear Relaxation

The fundamental equations (1.7) have a thermodynamic sense and show that a nuclear system can be excited by radiofrequency irradiation to absorb the energy. If a new excited state of nuclei is not isolated and capable of an energy exchange with the environment, the initial equilibrium state can be recovered. There are two principally different relaxation mechanisms. The first one corresponds to the energetic exchange between excited nuclear *spins* and the *lattice*. The latter can be formulated as a continuum of nuclear magnetic moments (spins) of any sort, which surround the nuclear spins detected in NMR experiments. In chemical language, the spins, populating the lattice, can be physically located in the same molecule or neighboring molecules and solvent molecules. By definition, spin–lattice nuclear relaxation reduces the total energy of the excited nuclear system, leading to recovery of the equilibrium state. In contrast, the second relaxation channel has rather an entropy character: two *identical* spins with different orientations with respect to the external magnetic field undergo motions 'flip–flop' as shown in Figure 1.8. It is obvious that the total energy of the nuclear system remains

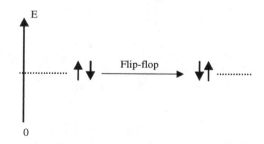

Figure 1.8 Flip–flop spin motions causing nuclear spin–spin relaxation

unchanged. Nevertheless, under these conditions, lifetimes of the excited spins decrease and thus, the nuclear system relaxes.

1.2.1 Macroscopic Magnetization: Relaxation Times T_1 and T_2

The spin–lattice and spin–spin relaxation mechanisms can be clearly imagined in terms of macroscopic magnetization. The latter appears in the sample when it is placed in the external magnetic field. In the equilibrium state, the macroscopic magnetization vector, $M^0{}_Z$, is lying along the direction of the applied magnetic field, the Z-direction in the coordinate system of Figure 1.9. In this state, the $M^0{}_X$ and $M^0{}_Y$ component of the magnetization vector are equal to zero. Note that the creation of the macroscopic magnetization is directly related to the behavior of nuclear dipoles placed in the external magnetic field. The nuclear dipoles with $m_I = +\frac{1}{2}$ and $-\frac{1}{2}$ (Figure 1.10)

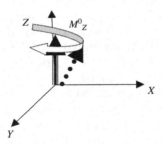

Figure 1.9 Macroscopic magnetization vector before and after action of a radiofrequency pulse. After the excitation, the macroscopic magnetization vector M_Z deviates from the Z-axis and precesses with the Larmor frequency

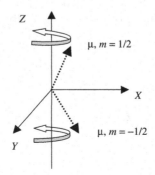

Figure 1.10 Precession of nuclear magnetic moments around the Z-axis. The external magnetic field is applied along this direction

precess around the Z-direction with the Larmor frequency. Since their orientations with respect to B_0 are energetically different and the orientation, corresponding to $m_I = +\frac{1}{2}$, is preferable, the dipoles are summarized to give finally the macroscopic magnetization vector along the Z-axis.

Irradiating the sample at the Larmor frequency ν_0 leads to reorientation of the $M^0{}_Z$ vector and its $M^0{}_Z$ projection reduces to M_Z (Figure 1.9). Under these conditions, the nuclear system is excited, the M_Z vector precesses around the Z-direction and the instantaneous M_X and M_Y components are nonzero. Bloch has described the relaxation of the excited nuclear system in terms of the magnetization vector located in a coordinate system, rotating at the frequency ν_0. In this coordinate system, the M_Z vector behaves according to the equation:

$$dM_Z/dt = -(M_Z - M_Z{}^0)/T_1 \tag{1.9}$$

and thus nuclear spin–lattice relaxation can be formulated as recovery of the Z-(*longitudinal*) component of the nuclear magnetization vector with time constant T_1. By analogy, spin–spin relaxation corresponds to the Y- and X- (*transverse*) components reducing to zero with time constant T_2:

$$dM_Y/dt = -M_Y/T_2$$
$$dM_X/dt = -M_X/T_2 \tag{1.10}$$

It is obvious that phenomenologically the T_1 and T_2 times characterize two different processes and therefore, in the common case, $T_1 \neq T_2$. In fact, $T_1 > T_2$ in solids where molecular motions are strongly restricted. In contrast, in solutions and liquids with fast molecular motions these times are usually very similar.

Bloch's description of nuclear relaxation provides an analytical means to deduce the shape of NMR signals as a narrow bell-shaped curve with a maximum at frequency ν_0 (Figure 1.11). It must be emphasized that this, so-called, Lorenz shape is typical of liquids and solutions, but not solids. When the external magnetic field has an ideal homogeneity, linewidths, $\Delta\nu$, measured in Hz, are directly controlled by T_1 and T_2 relaxation times according to:

$$\Delta\nu = 1/\pi\, T_{1,2} \tag{1.11}$$

However, in practice, the external magnetic field is not ideal and then the width of NMR signals depends on the so-called effective relaxation time $T^*{}_2$:

$$1/T^*{}_2 = (\gamma\Delta B_0/2) + 1/T_2 \tag{1.12}$$

It is seen that $T_2{}^* \to T_2$ when nonhomogeneity of the external magnetic field ΔB_0 is minimal. This situation, reached by standard adjustments of shimming coils in NMR spectrometers, leads to narrowing of the observed resonance lines and increase of their peak intensities. The latter, as we will

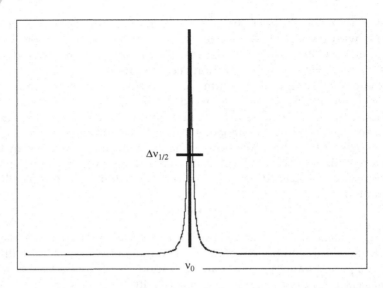

Figure 1.11 The Lorenz shape of an NMR signal, which can analytically deduced for liquids on the base of the Bloch equations

see below, is very important in NMR relaxation measurements. Note that in the case of diamagnetic solutions and non-quadrupolar nuclei the standard shimming adjustments usually give linewidths less than 0.3 Hz.

According to relationships (1.11) and (1.12), shortening relaxation T_1, T_2 times results in natural broadenings of NMR signals. These effects are particularly strong for nuclei with spins $I \geqslant 1$. As we will see below, such nuclei relax by the mechanism of quadrupolar interactions and T_1, T_2 times are actually short. To show practically the variations in the T_1 times and the expected line widths, consider the data in Table 1.2. For example, ^{1}H and ^{31}P nuclei ($I = \frac{1}{2}$) of compound $(c\text{-}CH_2O)_2P(BH_3)N(CH_3)_2$ (Figure 1.12) relax in solutions with the quite long T_1 times of 3.8 and 9.6 s, respectively.

Table 1.2 T_1 times of nuclear relaxation, measured in solutions of some compounds, and line widths Δv expected in NMR spectra

Nucleus	Spin	Compound	T_1 (s)	Δv (Hz)
^{2}H	1	CF_3COOD	0.015	21
^{11}B	3/2	$(c\text{-}CH_2O)_2P(BH_3)N(CH_3)_2$	0.021	15
^{1}H	1/2	$(c\text{-}CH_2O)_2P(BH_3)N(CH_3)_2$	3.8	0.08
^{31}P	1/2	$(c\text{-}CH_2O)_2P(BH_3)N(CH_3)_2$	9.6	0.03
^{14}N	1	Pyridine	0.0016	200
^{15}N	1/2	Pyridine	85	0.004
^{93}Nb	9/2	$CpNbH_3$	0.00002	16000

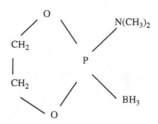

Figure 1.12   Dimethylaminephospholane, investigated by the ^{1}H, ^{11}B and ^{31}P NMR relaxation in an O_2-free benzene-d_6 solution (see Table 1.2)

In accordance with Equation (1.11), the lines in the ^{1}H and ^{31}P NMR spectra of this compound are narrow. In contrast, ^{11}B nuclei with $I = 3/2$ in the same compound relax much faster (^{11}B $T_1 = 21$ ms). As result, an expected line width in the ^{11}B NMR spectrum will be more than 15 Hz. It is obvious that such natural broadenings significantly exceed the effects that could be caused by nonhomogeneity of the external magnetic field. ^{15}N ($I = 1/2$) and ^{14}N ($I = 1$) isotopes in pyridine molecules (T_1 (^{15}N) and T_1 (^{14}N) times are measured as 85 and 0.0016 s, respectively) demonstrate the strongly greater effects. Finally, detection of NMR signals can become problematic or even impossible when relaxation times are extremely short. For example, an expected width of the ^{93}Nb NMR signal in compound $CpNbH_3$ is as large as 16 kHz. In such cases, it is difficult to distinguish resonance signals from baselines in NMR spectra.

1.3   Molecular Motions as the Reason for Nuclear Relaxation

Nuclear spin–lattice relaxation, as a physical phenomenon, is an energetic exchange between excited nuclear spins and their environment. The M_Z component of the macroscopic magnetization vector, turned from its equilibrium orientation (Figure 1.9), precesses around the Z-direction and thus the energetic exchange between the excited spins and lattice will be possible in principle if the lattice creates magnetic fields, *fluctuating* at frequencies close to the Larmor resonance frequency ν_0 [1]. Nuclear magnetic moments, forming the lattice, are physically located in molecules. In turn, the molecules undergo thermal motions, which lead to the appearance of the fluctuating fields responsible for thermal nuclear relaxation.

Figure 1.13 explains schematically how molecular motions can create such fluctuating fields. Magnetic dipoles of the A and B nuclei, located, for example, in the same molecule, are oriented along the direction of the applied external magnetic field B_0. These dipoles can interact and the strength of the

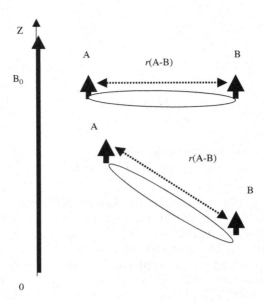

Figure 1.13 Schematic presentation of molecular reorientation on dipolar coupling between spins of A and B nuclei located in a molecule placed in the external magnetic field B_0. The molecular motion leads to a change in the orientation of the dipolar vector with respect to the direction of the external magnetic field

dipole–dipole A–B coupling depends on the internuclear distance, r(A–B), and spatial orientation of the dipolar vector, r(A–B), with respect to the Z-axis. Then, strength of a magnetic field, created by the A nuclear magnetic dipole at the location of the B nucleus also depends on the internuclear distance and the r(A–B) orientation. It is obvious now that if a molecular motion, shown in Figure 1.13, leads to reorientation of the r(A–B) dipolar vector, then the magnetic field on the B nucleus also changes. On the other hand, it easy to see that any molecular motion, for example, a free rotation around the axis, lying along the r(A–B) vector, does not change its spatial orientation and hence it has no influence on nuclear relaxation. Figure 1.14

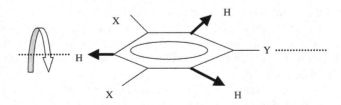

Figure 1.14 Molecular motions of trisubstituted aromatic molecules, having a preferable rotational axis, located along the H–Y vector

illustrates this situation in an aromatic molecule having a preferred rotational axis. It is seen that the rotation leads to reorientation of C−H bonds in *ortho*-positions of the aromatic ring while the direction of the C−H bond in the *para*-position does not change. For this reason, T_1 times of carbons, relaxing by the dipole–dipole carbon–proton mechanism and lying along the rotation axis, are usually remarkably longer.

Finally again it must be emphasized that molecular motions affect nuclear relaxation more when the frequencies of motions are close to the Larmor resonance frequency ν_0. That is why relaxation times in solids are usually significantly longer than those in liquids and solutions.

1.3.1　Correlation Times and Activation Energies of Molecular Motions

Molecules in liquids and solutions undergo fast thermal motions such as rotational reorientations, translational motions or their combinations. Additionally, various intramolecular motions (rotations around single chemical bonds and segmental motions in polymeric molecules etc.) can significantly contribute to nuclear relaxation. Quantitatively the molecular motions are characterized by the correlation times τ_C and the activation energies ΔE_a. The magnitude of τ_C can be formulated as the time necessary for a molecule to be reoriented (a tumbling time). It is obvious that any molecular reorientation, even in solutions, requires structural changes in the immediate environment. That is why the moving molecule should overcome an energy barrier. Then, by analogy with chemical kinetics, one can write:

$$\tau_C = \tau_0 \exp(E_a/RT) \tag{1.13}$$

where E_a is the activation energy of molecular motions and τ_0 is the correlation time constant. Again, by analogy with chemical kinetics, the magnitude of τ_0^{-1} can be defined as a frequency of attempts to overcome the energy barrier and therefore $\tau_0 \sim 10^{-13}–10^{-14}$ s. It is clear that a larger correlation time corresponds to a slower motion. Thus, on cooling, molecular reorientations slow down according to an exponential law. However, a more correct definition of the τ_C is connected with the, so-called autocorrelation function in the theory of nuclear relaxation where the τ_C is an average time for the molecule to progresses through one radian.

The Stokes–Einstein–Debye theory of the liquid state shows that the molecular motion correlation times depend on sizes (or volumes) of molecules and bulk viscosity of solvents η. In the case of rotational reorientations of spherical molecules (which behave like totally rigid bodies), the correlation time τ_C is written as:

$$\tau_C = 4\pi a^3 \eta/3kT \tag{1.14}$$

where a is the molecular radius. This relationship explains why the correlation times increase in solutions on going from small organic molecules

($\tau_C \leqslant 10^{-11}$ s) to polymeric systems ($\tau_C \geqslant 10^{-9}$ s). In practice, the rotational correlation time is given by:

$$\tau_C = (4\pi a^3 \eta / 3kT)\phi C + \tau_0 \qquad (1.15)$$

where ϕ is a shape parameter (1 for a sphere), τ_0 is the inertia contribution to the overall rerientational time. Usually in liquids, this term is ignored. C is the, so-called, slip coefficient, which can be calculated via several theories and measures the hindrance to rotation experienced by a molecule in a dense liquid [6].

Abragam [1] gives the best illustration of strong effects, caused by the viscosity of samples (Table 1.3). As we will demonstrate below, the proton T_1 relaxation time, determined at room temperature, is a measure of the correlation time τ_C because $1/T_1$ is proportional to τ_C. It follows from Table 1.3 that, increasing the viscosity leads to the shortening of ^{1}H T_1 times and hence to an increase of molecular motion correlation times. Finally, sometimes rotational diffusion constants D are used for characterization of molecular motions instead of correlation times. Equation (1.16) gives the relation between these.

$$D = 1/6\,\tau_C \qquad (1.16)$$

In solutions, rotational molecular reorientations play a major role. Activation energies of such motions in the case of non-aggregated molecules usually take values between 0.8 and 5 kcal/mol. Table 1.4 illustrates E_a values typical of middle-sized transition metal complexes. Comparison of the E_a values in toluene and CD_2Cl_2 for molecules $W(H_2)(CO)_3(PPri_3)_2$ or $OsH(H_2)(Cl)(CO)(PPri_3)_2$ demonstrates the influence of solvent viscosities. As can be seen, the more viscous medium (toluene) corresponds to a higher E_a value. The same effect is observed in $CDCl_3$ and toluene solutions of compound $(\mu\text{-}H)_2Os_3(CO)_{10}$. In accordance with Equation (1.14), activation energies for compounds with similar molecular volumes (for example, $OsH(H_2)(Cl)(CO)(PPr^i_3)_2$ and $OsH(O_2)(Cl)(CO)(PPr^i_3)_2$ or $[FeH(H_2)(dppe)_2]$ BF_4 and $[RuH(H_2)(dppe)_2]$ BF_4) are practically identical. Finally activation energies are maximal in the case of binuclear systems such as $Cp_2TlH_2H\text{-}W(CO)_5$ and $(MeCp)\{MeCp\text{-}PPh_2Cr(CO)_4\}W(\mu\text{-}H)H$.

Table 1.3 Proton spin–lattice relaxation times measured in different liquids [1]

Liquid	η (cP)	T_1 (s)
Petroleum ether	0.48	3.5
Ligroin	0.79	1.7
Kerosene	1.55	0.7
Heavy machine oil	260	0.013
Mineral oil	240	0.007

Table 1.4 Activation energies of molecular reorientations of transition metal hydride complexes, determined from variable-temperature ^{1}H NMR relaxation data in solution [7]

Compound	E_a (kcal/mol)	Solvent
$OsH_4(PTol_3)_3$	2.53	CD_2Cl_2
$OsH(H_2)(Cl)(CO)(PPr^i_3)_2$	2.6	CD_2Cl_2
$OsH(H_2)(Cl)(CO)(PPr^i_3)_2$	3.8	Toluene-D$_8$
$OsH(H_2)(Cl)(CO)(PPr^i_3)_2$	3.2	Toluene-D$_8$
$OsH(O_2)(Cl)(CO)(PPr^i_3)_2$	3.3	Toluene-D$_8$
$[FeH(H_2)(dppe)_2] BF_4$	2.6	Acetone-D$_6$
$[RuH(H_2)(dppe)_2] BF_4$	2.5	Acetone-D$_6$
$[OsH(H_2)(depe)_2] BPh_4$	2.9	Acetone-D$_6$
$HMn(CO)_5$	2.52	TDF
$HMn(CO)_3(PEt_3)_2$	2.8	CD_2Cl_2
$W(H_2)(CO)_3(PPr^i_3)_2$	2.6	CD_2Cl_2
$W(H_2)(CO)_3(PPr^i_3)_2$	3.8	Toluene-D$_8$
$Cp_2TlH_2H\text{-}W(CO)_5$	4.0	Toluene-D$_8$
$Re_2(\mu H_2)(CO)_8$	3.1	Toluene-D$_8$
$(MeCp)\{MeCp\text{-}PPh_2Cr(CO)_4\}W(\mu\text{-}H)H$	4.2	Toluene-D$_8$
$(\mu\text{-}H)_2Os_3(CO)_{10}$	2.2	$CDCl_3$
$(\mu\text{-}H)_2Os_3(CO)_{10}$	3.4	Toluene-D$_8$
$H(\mu\text{-}H)Os_3(CO)_{11}$	4.4	Toluene-D$_8$

Commonly, molecular motions in solids are much slower than in solutions. In the temperature region between $-150°$ and $+250°C$ their correlation times take values of 10^{-4}–10^{-6} s and activation energies reach 15–20 kcal/mol and more. Exceptions are small molecules, which can undergo fast motions, even in the solid state. For example, in the high-temperature tetragonal phase (above 225 K) solid BD_3ND_3 shows reorientations with the E_a values of 1.4 kcal/mol ($\tau_0 = 1.1 \times 10^{-13}$ s) and 1.7 kcal/mol ($\tau_0 = 4.4 \times 10^{-14}$ s) for BD_3 and ND_3 groups, respectively. In the low-temperature orthorhombic phase, these motions require higher energies: $E_a = 3.2$ and 6.2 kcal/mol [8].

1.3.2 Isotropic and Anisotropic Molecular Motions

A molecular motion is *isotropic* when a *single* value of the correlation time τ_C at a given temperature (and a *single* value of the activation energy E_a, respectively) describe this motion. For geometric reasons, the fully symmetric rigid spherical molecules (Figure 1.15) undergo isotropic rotational reorientations in dilute solutions. Note however that the ideal spherical molecules are rather rare. Nevertheless motions of various symmetric molecules, such as, the octahedral complexes $Re(CO)_6$ or $Mn(CO)_6$ can be satisfactorily described as isotropic.

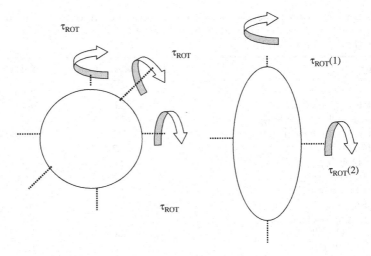

Figure 1.15 Isotropic rotational reorientations of the ideal spherical molecule (left) and anisotropic rotational motions of a molecule having a symmetrically-ellipsoidal shape (right)

By definition, a symmetric ellipsoid, shown in Figure 1.15, has two directions of spatial extension. For this reason, such a molecule has two different moments of inertia and thus rotational reorientations will be characterized by two correlation times, $\tau_C(1)$, $\tau_C(2)$, and two activation energy values. This simplest case of an *anisotropic* molecular motion often occurs in chemistry. However, in most cases, motions are more complicated, particularly for polymeric (and biological) molecules or bulky inorganic aggregates. Table 1.5 lists the effective motional correlation times, obtained for protonated carbons in the molecule, depicted in Figure 1.17, from the ^{13}C T_1

Table 1.5 ^{13}C T_1 times of protonated carbons in piroxicam (see Figure 1.17) and the effective molecular motion correlation times in DMSO-d_6 (295 K) [10]

Carbon	T_1 (s)	τ_C (s)
14	0.476	9.1×10^{-11}
12	0.476	9.1×10^{-11}
4	0.312	1.42×10^{-10}
3	0.25	1.8×10^{-10}
5	0.385	1.14×10^{-10}
2	0.385	1.14×10^{-10}
13	0.196	2.40×10^{-10}
11	0.500	8.70×10^{-10}

measurements in DMSO-d$_6$ solutions [10]. In spite of the fact that each carbon is attached to one proton, all the carbons show different relaxation times
and correlation times. The data reveal a complex character of molecular
motions where, beside the molecular tumbling, there are internal rotations
with a preferred axis (Figure 1.17). Even the simplest organic molecules can
undergo anisotropic motions. For example, molecular motions of toluene in
net liquid are composed of reorientations around the axes ZZ, YY and XX
(Figure 1.16). Table 1.6 lists the correlation times of these motions and their
activation energies. As we will see below, times of nuclear relaxation depend
strongly on rates and the character of molecular motions. The latter plays a
very important role when the relaxation times are interpreted quantitatively.

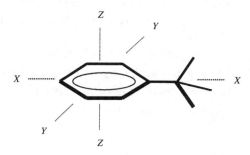

Figure 1.16 Molecular rotational reorientations of toluene in neat liquid

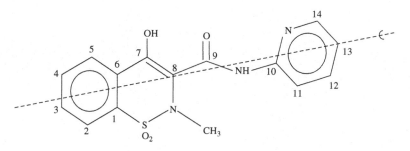

Figure 1.17 Structure, atom numbering, and the main axis of internal motion in
4-hydroxy-2-methyl-N-2-pyridinyl-2H-1,2-benzothiazine-3-carboxamide-1,2-dioxide.
(Reproduced with permission from C. Rossi, *Chemical Physics Letters* 1992; **193**, 553)

Table 1.6   Parameters of anisotropic motions of toluene
molecules in neat liquid [9]

Axis	τ_C at 295 K (s)	τ_0 (s)	E_a (kcal/mol)
XX	0.98×10^{-11}	2.5×10^{-12}	0.8
YY	2.2×10^{-11}	2.9×10^{-12}	1.2
ZZ	1.1×10^{-11}	0.25×10^{-12}	2.2

Bibliography

1. A. Abragam. *Principles of Nuclear Magnetism*. Oxford at the Clarendon Press: New York, 1985.

2. C. T. Farrar. *An introduction to pulse NMR spectroscopy*. Farragut Press: Chicago, 1987.

3. H. Friebolin. *Basic One- and Two-Dimensional NMR Spectroscopy*. VCH: Weinheim, 1991.

4. J. B. Lambert, F. G. Riddel. *The Multinuclear Approach to NMR Spectroscopy*. Reidel: Boston, 1982.

5. R. K. Harris. *Nuclear Magnetic Resonance Spectroscopy*. Bath Press: Avon, 1983.

6. N. D. Martin, M. H. Issa, R. A. McIntyre, A. A. Rodriguez. *Journal of Physical Chemistry A* 2000; **104**: 11278.

7. V. I. Bakhmutov, E. V. Vorontsov. *Reviews of Inorganic Chemistry* 1998; **18**: 183.

8. G. H. Penner, Y. C. Chang, J. Hutzal. *Inorganic Chemistry* 1999; **38**: 2868.

9. L. Sturz, A. Dolle. *Journal of Physical Chemistry A* 2001; **105**: 5055.

10. C. Rossi. *Chemical Physics Letters* 1992; **193**: 553.

2 How to Measure NMR Relaxation Times

The hardware and software of modern commercial FT NMR spectrometers provide simple and convenient procedures for measurement of relaxation times. In this chapter we are concerned only with the methods of direct relaxation measurements suitable for target nuclei having a high natural abundance (^{1}H, ^{19}F, ^{31}P, ^{11}B etc.) and moderately long relaxation. In addition, these methods are particularly successful when compounds under investigation are relatively simple, giving well-resolved NMR spectra. Indirect relaxation measurements, addressed to studies of complex molecules and based on application of two-dimensional techniques, often require a special design of NMR probes. These methods will be considered in Chapter 11.

The spin–lattice and spin–spin relaxation can be probed by the standard inversion recovery and Carr–Purcell–Meiboom–Gill pulse sequences. In most cases, quantitative structural information is extracted from an analysis of the spin–lattice relaxation times. Therefore, here we focus mainly on methods for collection of T_1 data.

Experiments in a rotating coordinate system, based on the spin locking pulse sequence, lead to determinations of $T_{1\rho}$ times. As we will show, $T_{1\rho}$ relaxation times are closely related to times of spin–spin nuclear relaxation [1]. Moreover, $T_{1\rho}$ and T_2 times, measured at moderate temperatures, are identical for most liquids and solutions. In the context of structural applications, T_2 and $T_{1\rho}$ data are less informative. At the same time, they are actively applied for studies of molecular mobility. These aspects will be considered also in Chapter 11.

Practical NMR Relaxation for Chemists   Vladimir I. Bakhmutov
© 2004 John Wiley & Sons, Ltd   ISBNs: 0-470-09445-1 (HB); 0-470-09446-X (PB)

2.1 Exponential and Non-exponential Nuclear Relaxation

As we have seen, nuclear relaxation, as recovery of an equilibrium state, is a time-dependent process described by Equations (1.9) and (1.10). Integration of Equation (1.9) with $M_Z = -M_Z{}^0$ at $t = 0$ results in the equations:

$$\ln(M^0{}_Z - M_z) = \ln 2M^0 - \tau/T_1 \qquad (2.1)$$

$$M_Z/M_0 = 1 - 2\exp(-\tau/T_1) \qquad (2.2)$$

showing an *exponential* character of the relaxation. Theoretically, nuclear relaxation is simple and *monoexponential* only for *uncoupled* (i.e. isolated) or *weakly coupled* nuclear spins. In the context of structural applications, this type of nuclear relaxation is mostly 'desirable' because the free induction decays, collected experimentally, can be automatically treated with the help of the standard software of NMR spectrometers to give finally T_1 (or T_2) relaxation times. However, the *strongly anisotropic* molecular motions can lead to a *nonexponential* spin–lattice relaxation, even for simple spin systems and in solutions. Registration, analysis and treatment of such a relaxation can be found elsewhere [2]. The *strongly coupled* spin systems, usually formed by ^{1}H nuclei, often show the complex relaxation behavior, expressed in practice as the *multiexponential* free induction decays. The principal feature of these decays is the fact that they cannot be treated and described by a *single* time parameter T_1 (or T_2). Calculations of the multiexponential free induction decays are not trivial [3]. In addition, they require input from certain relaxation models based on independent data. In the context of structural applications this type of *strongly-coupled* nuclear relaxation is 'undesirable'.

2.2 Measurements of Spin–Lattice Relaxation Times

T_1 relaxation time measurement can be carried out by *inversion recovery*, *saturation recovery* or *progressive saturation* NMR experiments. Other methods of direct T_1 determination are also available [4]. However, for reasons of simplicity and convenience the inversion recovery approach remains the most popular. Relaxation measurements are based on applications of radiofrequency pulse sequences where the operating pulses can have different power and length. Variations in these pulse parameters can excite either the full NMR spectrum or only part of it.

 Let us introduce the definition of the acting angle of a radiofrequency pulse. When the macroscopic magnetization is considered in a coordinate system,

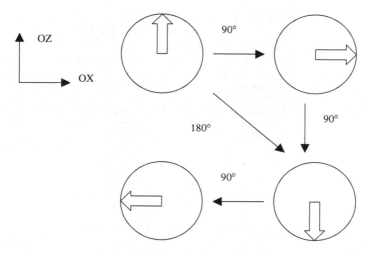

Figure 2.1 Schematic presentation of the behavior of the macroscopic magnetization vector M in a rotating coordinate system after action of a 90°, 180° radiofrequency pulses or two consecutive 90° radiofrequency pulses. The M orientation along the positive X-axis corresponds to maximal intensity of a registered NMR signal

rotating with the Larmor frequency, the equilibrium state of a nuclear system corresponds to the location of the magnetization vector along the Z-direction in Figure 2.1. An irradiating field, exciting the nuclei, is applied, as a short radiofrequency pulse, along the Y-axis (i.e. perpendicular to the plane of Figure 2.1) and converts the magnetization vector to the X-axis. Since NMR signals are technically registered along this direction, the acting angle α of the applied radiofrequency pulse, is defined as:

$$\alpha = \gamma B_1 t_p$$

$$\cos(\alpha) = \exp(-t_p/T_1) \tag{2.3}$$

where B_1 is the power of the pulse and t_p is its duration. In these terms, by definition from Figure 2.1, a 90° pulse will generate a maximal magnetization along the X-axis. Then, after Fourier transformation, line intensities in the recorded NMR spectrum are maximal. In contrast, a 180° pulse converts the magnetization vector from Z to $-Z$ and its projection on the X-axis is equal to zero. In such case, after Fourier transformation, line intensities will be also close to zero. It follows from Figure 2.1 that pulses <90° or >90° do not lead to optimal line intensities in resulting NMR spectra. Thus, the length of the pulse plays an important role in long-term NMR experiments with a great number of accumulations. An optimal length, providing the best sensitivity for a given relaxation time, is called the Ernst angle. When NMR data are collected with the help of the one pulse sequence, shown in Figure 1.2, and

Table 2.1 The Ernst angle α providing the best sensitivity in NMR spectra accumulated in long-term experiments with one pulse sequence at zero relaxation delay and acquisition time 1 s

Spin–lattice relaxation time (s)	Ernst angle α (°)
100	8
10	25
4	33
2	53
1	68
0.4	86
0.1	90

the RD value is zero, then nuclei relax during the acquisition time AT. At the AT value of 1 s, the Ernst angle, as a function of T_1, can be found with the help of the data in Table 2.1. At long relaxation times (100 s) the resulting NMR signals will be accumulated more successfully by using the 8° pulses while 90° pulses are more productive at the short relaxation times.

The standard inversion recovery experiments for determinations of the *non-selective* T_1 times are based on the pulse sequence:

$$(RD-180°-\tau-90° - AT)_n \tag{2.4}$$

where n is a number of accumulations. Figure 2.2 schematically illustrates the action of the pulse sequence and the resulting NMR spectrum, obtained after Fourier transformation. Here the time delay τ between the inverting and registering pulses is varying from $\tau \ll T_1$ to $\tau \simeq 3T_1$. It is obvious that at a very short τ delay ($\tau \ll T_1$) the first 180° pulse inverts the magnetization vector to the Z-axis and the registering 90° pulse converts this vector from $-Z$ to $-X$. As a result, a NMR signal appears with a 'negative' intensity.

All the pulses in the inversion recovery sequence (2.4) are *hard* and *short*. Note that the hard pulses have typical durations of $\sim$5–10 μs and excite a range of frequencies of order 10^5 Hz. Thus, the hard pulses excite all resonances in the NMR spectra and for this reason, the determined relaxation times are *nonselective*.

The NMR spectra, obtained at τ variations in the inversion recovery experiments, will show their evolution from the negative intensities to the positive ones (Figure 2.3). The collected data can be treated by a standard nonlinear three-parameter fitting routine of NMR spectrometers to calculate T_1 values. The three parameters (an initial intensity, a current intensity and a time T_1) are varied to minimize differences between the experimental and calculated line intensities by a least-squares method. Figure 2.3 illustrates the

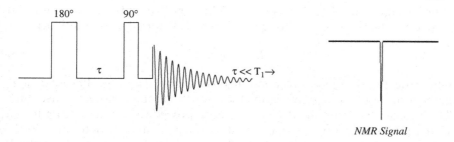

Figure 2.2 Inversion recovery experiments for T_1 determinations. The resulting NMR spectrum is shown for a very short τ time between the inverting and registering pulses

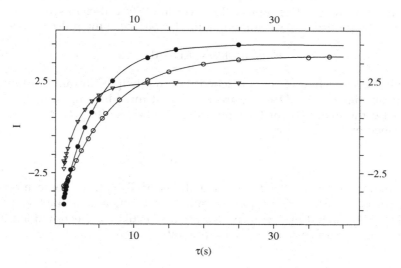

Figure 2.3 Inversion recovery curves collected experimentally (points) and calculated with the standard nonlinear three-parameter fitting routine (solid lines). The data are collected for CH_3 (triangles) and α-protons (solid circles) in compound $(CH_3)(BF_3)N(CH_2CH_2)_2$ ($CDCl_3$, 25°C, $T_1 = 4.8$ and 5.1 s, respectively) and α-protons (open circles) in $(H)(BF_3)N(CH_2CH_2)_2$ (CD_2Cl_2, 25°C, $T_1 = 4.8$ s). (Reproduced with permission from M. Güizado-Rodríguez *et al. Inorganic Chemistry*, 2001; **40**: 3243. © 2001 American Chemical Society)

typical example of such treatments where the solid lines correspond to results of the fitting procedures [5]. Again, the standard programs, provided by the software of commercial NMR spectrometers, fit the experimental points to *monoexponential* curves and they cannot be applied for more complicated cases. Finally, to improve the quality of spin inversion [6], the 180° pulse in sequence (2.4) can be replaced with a composite pulse cluster 90°$_\phi$ 240°$_\phi$ 90°$_\phi$.

From Figure 2.3, line intensities in NMR spectra, obtained by inversion recovery experiments, go through zero values. Under these conditions:

$$T_1 = \tau_0 / \ln 2 \qquad (2.5)$$

where the time delay τ_0 corresponds to observation of minimal line intensities in the resulting NMR spectra or, in other words, to zero crossing points. Thus, T_1 times can be determined without fitting procedures directly from the τ_0 values [1]. However, it is obvious that accuracy in such T_1 determination is not very high. In fact, the resulting T_1 values, determined by this method, will depend strongly on signal/noise ratios in the collected NMR spectra.

Pulse sequences, applied for T_1 determinations by saturation recovery experiments, are similar to the inversion recovery sequences. Here however the 90° saturating pulses replace the inverting 180° pulses in sequence (2.4). Finally, the progressive saturation method is based on the pulse sequence:

$$Dummy\ pulses-RD-90°-AT \qquad (2.6)$$

where the saturation effect is reached with the help of dummy pulses and the delay times, $\tau = RD + AT$, are varying. It must be emphasized that in this case the evolution of line intensities in the resulting NMR spectra is described by the equation:

$$M^\tau_Z / M_Z^0 = 1 - \exp(-\tau/T_1) \qquad (2.7)$$

This equation can be used for calculations of T_1 values. Some modified inversion recovery versions (the so-called fast methods) can be taken from Kingsley [4], including the formulas for T_1 calculations, estimations of T_1 errors and maximization of signal/noise ratios.

2.3 Measurements of Selective and Biselective T_1 Times

The duration and the power of 180° pulses in sequence (2.4) can be adjusted to excite a *single line* in an NMR spectrum. Action of the *selective* (or *soft*) 180° pulses in the inverse recovery experiments:

$$RD-180°_{sel}-\tau-90°-AT \qquad (2.8)$$

leads to determination of the *selective* relaxation times T_{1sel}. At short τ values ($\tau \ll T_1$) the resulting NMR spectrum will show the single resonance with a 'negative intensity' (Figure 2.4). It is obvious that these conditions can be used for calibration of the soft 180° pulses: a good calibration corresponds to

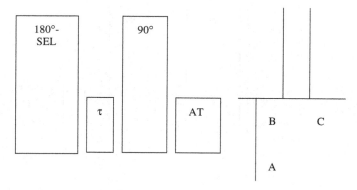

Figure 2.4 Inversion recovery experiment with the selective 180° inverting pulse operating in the region of the A resonance. Typical durations of selective pulses are 30–40 ms. The resulting NMR spectrum is shown for a very short τ time between the inverting and registering pulses

a maximally possible negative intensity of the excited resonance, which is, maximally close to its natural intensity.

In practice, the selective T_1 experiments are important in studies carried out at proton frequencies. Here a channel of ^1H decoupler can be used to generate the selective 180° pulses. Typical durations of these pulses are usually close to 20–30 ms. Finally, principal channels of NMR spectrometers are also suitable to obtain the selective pulses. Note that in these cases, they use the, so-called shaped pulses.

By analogy, pulse sequence (2.9) containing two soft inverting 180° pulses:

$$RD-180°_{sel}-180°_{sel}-\tau-90°-AT \qquad (2.9)$$

and operating at two different frequencies (Figure 2.5), results in determination of the *biselective* relaxation times T_{1bis}. Interpretation of these experiments as well as the T_{1sel} measurements is mostly successful in the case of protons. Their practical applications will be considered below. Here we emphasize that in most cases the ^1H relaxation times decrease as:

$$T_{1sel} > T_{1bis} > T_1 \qquad (2.10)$$

However, even theoretically, the times differ slightly. The differences are particularly small when two or more relaxation mechanisms operate simultaneously [7]. Figure 2.6 summarizes the results of the ^1H T_1, ^1H T_{1sel} and ^1H T_{1bis} experiments, carried out on the HX resonance of the Nb hydride complex in a toluene solution at room temperature [8]. The differences in the selective, biselective and nonselective $T_1(H_X)$ times are very insignificant because the spin–lattice relaxation of hydride resonances in such systems

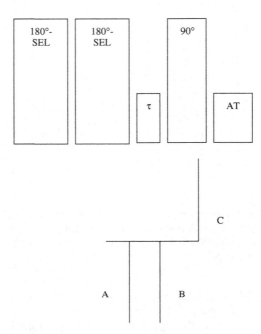

Figure 2.5 Inversion recovery experiment with two selective 180° inverting pulses operating at frequencies of the A and B resonances. The resulting NMR spectrum is shown for a very short τ time between the inverting and registering pulses

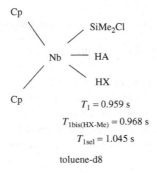

$$T_1 = 0.959 \text{ s}$$
$$T_{1\text{bis(HX-Me)}} = 0.968 \text{ s}$$
$$T_{1\text{sel}} = 1.045 \text{ s}$$

toluene-d8

Figure 2.6 Selective ($T_{1\text{sel}}$), biselective ($T_{1\text{bis}}$) and non-selective (T_1) times measured for the H_X hydride ligand in a toluene solution of $Cp_2NbH_2SiMe_2Cl$ at 400 MHz. The biselective experiments have been carried out for the H_X and CH_3 resonances

is governed by proton–proton and proton–niobium dipole–dipole inter-actions. It is obvious that interpretations of these measurements require a good accuracy in T_1 determinations. To obtain reliable data, three to five independent measurements are necessary.

2.4 Determination of $T_{1\rho}$ and T_2

As we discussed above, the action of the 90° pulse causes reorientation of the magnetization vector, as shown in Figures 1.9 and 2.1. When the irradiating field is switched off, the excited spins relax by molecular motions with frequencies close to the Larmor frequency. Figure 2.7 demonstrates another scheme of relaxation experiments where the irradiating radiofrequency field changes its phase by 90°, but it still operates as long as the τ time. Under these conditions, the nuclear magnetization is aligned along the direction of the applied radiofrequency field or, in other words, the magnetization vector is spin-locked by this effective field. The effective radiofrequency field is much weaker than the external magnetic field B_0. In most cases, the strength of the locking field is smaller by several orders of magnitude. For this reason, the created magnetization will decay with a specific time constant $T_{1\rho}$, governed by molecular motions with significantly lower frequencies. By analogy with the inversion recovery experiments, varying the τ values leads to determination of the spin–lattice relaxation time in the rotating coordinate system:

$$M(\tau) = M_0 \exp(-\tau/T_{1\rho}) \tag{2.11}$$

In the solid state, where molecular motions are strongly restricted, the $T_{1\rho}$ values differ remarkably from T_1 and T_2 times. That is why the $T_{1\rho}$ experiments are successfully applied for characterizations of molecular motions in solids. In contrast, molecular tumbling is fast in solutions and nonviscous liquids and therefore the T_1, T_2 and $T_{1\rho}$ times are practically identical.

The spin–spin relaxation time T_2 describes recovery of transverse components of the total nuclear magnetization. This time constant can be determined with the help of the Carr–Purcell pulse sequence:

$$(90°x'-\tau-180°y'-\tau \text{ (first echo)}-\tau-180°y'-\tau \text{ (second echo) etc.} \tag{2.12}$$

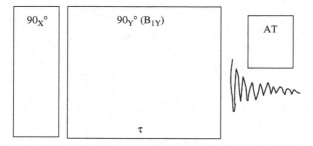

Figure 2.7 Spin-locking experiment for determination of $T_{1\rho}$ times. The radiofrequency field B_1, acts along the Y-axis and the time of its action (τ) is varied

By analogy with the inversion recovery experiments, the NMR decays, collected at τ variations, are treated by the standard fitting routine of NMR spectrometers to calculate T_2 times.

2.5 Preparation of Samples for Relaxation Experiments

Durations of relaxation times depend strongly on the nature of nuclei, relaxation mechanisms, structural features of molecular systems, their mobility, temperature and viscosity of solvents. By action of all these factors, relaxation times vary over very large ranges. Figure 2.8 illustrates the order and typical variations of proton T_1 times measured in solutions of simple organic molecules. As seen, the ^1H T_1 relaxation times change from 2.5 to 6 s in $CDCl_3$ and from 4.0 to 13 s in the less viscous CD_2Cl_2 [5]. T_1 times of ^{13}C nuclei vary much stronger. For example, ^{13}C nuclei directly attached to protons (CH, CH_2 and CH_3 groups) relax mainly by carbon–proton dipole–dipole interactions. For this reason, their T_1 times are relatively short (3–15 s). In contrast, quaternary carbons show very long T_1 times reaching 100 s and more. For example, ^{13}C T_1 times of fullerene C_{60} in various organic solvents are may be >120 s.

As we will show in Chapter 12, an unpaired electron causes a fast *paramagnetic* relaxation of neighboring nuclei. Since the unpaired electron spin has a magnetic moment, which is 10^3 times higher than nuclear magnetic moments, and since the paramagnetic relaxation rate is proportional to the squares of the electron and nuclear magnetic moments, the relaxation mechanism in the presence of unpaired electrons is very effective. For this reason, any paramagnetic impurities (such as simple paramagnetic ions and organic or organometallic species with unpaired electrons or even *molecular oxygen* in

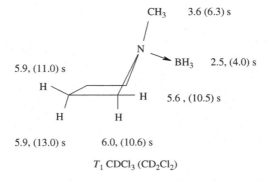

Figure 2.8 ^1H T_1 times measured in $CDCl_3$ and CD_2Cl_2 solutions (298 K, 400 MHz)

solvents) will reduce relaxation times of compounds under investigation. By definition, relaxation rates are additive magnitudes. Therefore, the influence of the paramagnetic contribution to a total relaxation rate depends on the duration of natural nuclear relaxation. It is obvious that a longer natural T_1 time will be distorted more. To avoid these effects, investigated samples should be previously purified and *deoxygenated*. The deoxygenation can be successfully reached by *four freeze–pump–thaw cycles*. The samples can be sealed under vacuum or inert atmosphere by conventional procedures if necessary. Good examples of the strong influence of oxygen are proton spin–lattice relaxation times of benzene molecules in solution. For example, a CS_2 solution of benzene in the presence of air shows the ^1H T_1 value of 2.7 s composed with 60 s in a degassed solution.

Bibliography

1. C. T. Farrar. *An Introduction to Pulse NMR spectroscopy*. Farragut Press: Chicago, 1987.

2. A. Kratochwill, R. L. Vold, J. Vold. *Journal of Chemical Physics* 1979; **71**: 1319.

3. D. M. Grant, C. L. Mayne, F. Liu, T. X. Xiang. *Chemical Reviews* 1991; **91**: 1591.

4. (a) P. B. Kingsley. *Concepts in Magnetic Resonance* 1999; **11**: 29; (b) V. I. Bakhmutov, E. V. Vorontsov. *Reviews of Inorganic Chemistry* 1998; **18**: 183.

5. M. Güizado-Rodríguez, A. Flores-Parra, S. A. Sánchez-Ruiz, R. Tapia-Benavides, R. Contreras, V. I. Bakhmutov. *Inorganic Chemistry* 2001; **40**: 3243.

6. K. Elbayed, D. Canet, J. Brondeau. *Molecular Physics* 1989; **68**: 295.

7. D. G. Gusev, D. Nietlispach, A. B. Vymenits, V. I. Bakhmutov, H. Berke. *Inorganic Chemistry* 1993; **32**: 3270.

8. V. I. Bakhmutov, J. A. K. Howard, D. A. Kenn. *Journal of the Chemical Society Dalton Transactions* 2000; 1631.

3 Errors in Determinations of Relaxation Times

The reliability of quantitative interpretations of relaxation times in terms of molecular structure or molecular mobility requires knowledge of errors in T_1 and T_2 determinations. Normally, potential sources of the errors are: (i) poor adjustments of NMR spectrometers, inaccurate measurements of the temperature and its poor stability (instrumental errors); (ii) a bad choice of varied parameters for T_1 (or T_2) measurements; (iii) incorrect treatments of the collected NMR decays, for example, when an inadequate relaxation model is applied for T_1 (or T_2) calculations; (iiii) the chemical nature of investigated samples. The latter can lead to a strong distortion of natural relaxation times, due to the presence of chemical exchanges or paramagnetics etc. Another source of the errors in interpretations of T_1 (or T_2) times is the nature of nuclear relaxation. Actually, many nuclei often relax by several mechanisms simultaneously. For example, the dipole–dipole relaxation and the relaxation by chemical shift anisotropy interactions or a set of different dipole–dipole interactions can contribute to magnitudes measured experimentally. In such cases, for example, calculations of structural parameters from T_1 data will be based on preliminary evaluations of the required contributions. It is obvious that incorrect evaluations of these contributions will lead to incorrect conclusions. It should be emphasized that such errors depend on concrete situations and they cannot be predicted in most cases.

Practical NMR Relaxation for Chemists Vladimir I. Bakhmutov
© 2004 John Wiley & Sons, Ltd   ISBNs: 0-470-09445-1 (HB); 0-470-09446-X (PB)

3.1 **Instrumental Errors**

The following instrumental factors have a significant influence on accuracy in relaxation time measurements:

(a) calibrations of radiofrequency pulses;

(b) homogeneity of radiofrequency pulses;

(c) temperature control of the relaxation experiments;

(d) signal/noise ratios in the resulting NMR spectra.

The latter factor is quite obvious because the calculations of relaxation times are based on treatments of peak (or integral) intensities of NMR signals collected in the relaxation experiments (Figure 3.1). Poor signal/noise ratios (particularly in the τ regions when the line intensities are close to zero) increase the errors. Factor (b) is rather a technical problem connected with the design of NMR spectrometers and does not depend on the investigator. The temperature of the NMR experiment is usually controlled with an accuracy of $\pm 0.5°C$ by standard variable-temperature units of commercial

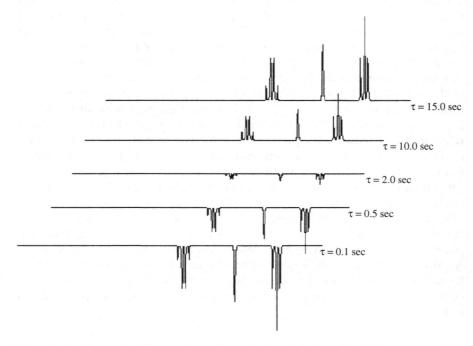

Figure 3.1 Schematic presentation of a typical evolution of line intensities in ^1H NMR spectra, collected by inversion recovery experiments at different values of τ

NMR spectrometers. Calibrations of a real temperature in a sample should be performed previously, for example, by the methanol thermometer method.

Good calibrations of the 90° (and 180°) radiofrequency pulses can be reached by standard single-pulse experiments with variations in the pulse durations t_p. In most cases, the procedures consist of searches of minimal line intensities in NMR spectra when the durations of the radiofrequency pulses correspond to the action angle α, close to 360°. Then durations of the 90° pulses are easy calculated as $t_p(360°)/4$. It should be emphasized that badly calibrated pulses often yield an effective *nonexponential* relaxation. In such cases, treatments of the collected free induction decays with the help of the standard fitting routine of NMR spectrometers will lead to additional errors in T_1 (or T_2) calculations. Finally, note that the calibrated durations of the radiofrequency pulses can remarkably change with temperature. Therefore, the calibration experiments should be carried out at various temperatures.

3.2 Incorrect Parameters for T_1, T_2 Measurements and T_1, T_2 Calculations

By definition, time delays RD in the pulse sequences, applied for relaxation experiments, should provide a *complete* relaxation of nuclei in each cycle of the measurements. It is obvious that if RD delays are too short, the peak (or integral) line intensities in the resulting NMR spectra will be distorted. In the case of a monoexponential nuclear relaxation, it is easy to show that the time delays, RD, close to T_1 values, recover ~63% of an equilibrium nuclear magnetization. In turn, a 99% recovery of the equilibrium magnetization will require time delays equal to 5 T_1. Therefore the T_1 measurements are correct when

$$\text{Relaxation delay} + \text{acquisition time} \geqslant 5T_1.$$

It is obvious that this condition requires preliminary NMR experiments for rough estimations of relaxation times. Since T_1, T_2 calculations are based on peak (or integral) line intensities, a bad phasing in the collected NMR spectra produces additional errors in T_1 or T_2 calculations. To minimize these errors, the final NMR spectra should be carefully phased by convenient procedures.

Intensity of lines I in the inversion recovery NMR spectra is a function of time τ: $I(\tau) = \exp(-\tau/T_1)$. It is obvious that semilogarithmic plots of $I-\tau$ will be straight lines with slopes equal to $1/T_1$. Determination of the slopes is the simplest method for T_1 calculations. Daragan *et al.* [1] have demonstrated that, owing to the presence of spectral noise, errors in determinations of the ln(I) values depend on the τ durations. Figure 3.2 illustrates schematically the resulting effect. As can be seen, the points, obtained at the *tails* of relaxation

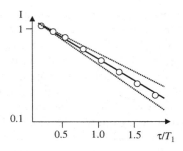

Figure 3.2 Line intensity plotted against τ/T_1 values in semilogarithmic (points and solid line). The dashed lines illustrate increasing errors due to the presence of noise in the NMR spectra

curves (when the τ/T_1 ratio is larger than one), give growing errors. It is obvious that these points should be weighted less. One can formulate the best region for the τ variation in T_1, T_2 experiments. In the case of a simple monoexponential relaxation, this region should be between $0.1T_1$ and T_1 where the number of the varying τ values is more than 16.

On the other hand in practice, the type of nuclear relaxation is often unknown. Therefore, in order to characterize properly, for example, the inversion recovery process, the τ values should cover a larger range from $0.1T_1$ to $3T_1$. This range allows one to distinguish exponential and nonexponential relaxation as systematic deviations of experimental points from the exponential behavior. For example, appearance of *overshoots* in the inversion recovery curves, particularly at τ values close to $T_1 - 2T_1$, is a good test for the presence of an unusual nuclear relaxation. Finally it should be noted that in spite of all these difficulties, the standard NMR adjustments and some experience in relaxation experiments lead to results where relaxation times are determined with errors less than 3–5%.

3.3 Coupled Nuclear Relaxation

By definition, two magnetically nonequivalent spins, A and X, form a simple system, AX, if the chemical shift difference between A and X is significantly larger than the spin–spin coupling constant J(A-X): $\Delta\delta^{AX}(Hz)/J^{AX}(Hz) > 10$. According to the well-known rule, the line multiplicity N of the A resonance is calculated as $N = 2nI + 1$, where n is the number of X nuclei and I is their spin. When the A, X nuclei are protons ($I = 1/2$), the AX spin system shows a ^1H NMR spectrum containing two doublets. Relative intensities within multiplets are distributed according to Pascal's triangle. Therefore, the AX system shows a 1:1 equilibrium distribution of line intensities in each doublet. Decreasing the $\Delta\delta^{AX}(Hz)/J^{AX}(Hz)$ ratio results in transformation of the AX spin system to a *strongly coupled* system, AB. By analogy, simple AX_2 or AXX' systems, formed by three nuclei, convert to strongly coupled systems marked as ABX, AMX, AB_2 or ABC. These systems show complex patterns where multiplicities of resonances and equilibrium intensities within multiplets do not follow the above rule or the binomial distribution. In each case the patterns depend on $\Delta\delta$ and J values.

As it has been mentioned, NMR relaxation of isolated spins is monoexponential. In contrast, strongly coupled spins can show 'coupled relaxation'. Ernst *et al.* have theoretically demonstrated that the registering hard 90° pulses, acting on strongly coupled spin systems in the inversion recovery experiments, cause a mixing of the eigenstate populations. In turn, this phenomenon is responsible for the appearance of *multiexponential* NMR decays. Schaublin *et al.* [2] have emphasized that such a relaxation can be observed, even in the case of a simple spin system AX_2. In practice, the presence of coupled nuclear relaxation can be experimentally recognized as a *strong perturbation of equilibrium distributions* in line intensities within resonance multiplets when an inverted spin system relaxes to equilibrium. Figure 3.3 illustrates this relaxation behavior for a simple AX_2 spin system. The distribution of line intensities in the A triplet does not correspond to 1:2:1 and changes during the relaxation. Standard treatments of these inversion recovery data will result in a faster effective relaxation of the central component of the A triplet. This effect is observed, for example, in the ^1H NMR spectra of the bimetallic trihydride complex (Figure 3.4), where the ^1H T_1 difference between the central and side lines in the H(A) triplet reaches 40% (Figure 3.5) [3].

It is obvious that treatments of the multiexponential NMR decays with the help of standard exponential fitting will give significant errors in relaxation

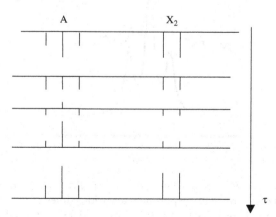

Figure 3.3 Coupled spin–lattice relaxation: the evolution of line intensities in an inversion recovery experiment on an AX_2 spin system (schematic)

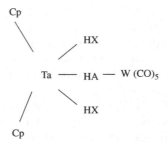

Figure 3.4 The structure of the bimetallic Ta/W trihydride complex, containing the bridging (H(A)) and terminal (H(X)) ligands. The H(A) and H(X) nuclei form an AX_2 spin system, showing the coupled proton relaxation in solutions

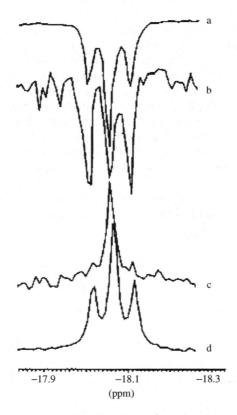

Figure 3.5 Room-temperature partially relaxed ^1H NMR spectra of the H(A) ligand collected in an inversion recovery experiment at various times τ for an acetone-d_6 solution of the complex in Figure 3.4: (a) τ = 0.1 s; (b) τ = 0.6 s; (c) τ = 1.1 s; (d) τ = 20.1 s. (Reproduced with permission from V. I. Bakhmutov, E. V. Vorontsov, G. Boni, C. Moise. *Inorganic Chemistry*, 1997; **36**: 4055. © 1997 American Chemical Society)

time calculations. Actually, such a formal treatment of the data, shown in Figure 3.5, gives $T_1 = 2.33$, 1.68 and 2.16 s for the lines in the H(A) triplet. Approaches to a quantitative analysis of the multiexponential NMR decays are based on perturbations of separated lines in spin multiplets, measurements of the return of the lines to an equilibrium state with further density matrix treatments [4]. Note that these studies rather concern description of molecular motions. In the context of structural applications, it must be emphasized that even the coupled relaxation can be successfully approximated by a single effective T_1 time when the equilibrium line intensities are *minimally perturbed* in the collected spectra [4].

3.4 Chemical Exchange

Relaxation times can be distorted in the presence of chemical (or positional) exchanges. By definition, a chemical exchange, for example, between two nonequivalent spin states, A and X,

$$\underset{\nu^A;\, \tau^A}{A} \quad \Longleftrightarrow \quad \underset{\nu^X;\, \tau^X}{X} \tag{3.1}$$

directly affects lineshapes of the A, X resonances when the exchange frequencies, ν_E, are comparable to chemical shift differences, $(\nu^A - \nu^X)$. Under these conditions, the NMR spectra show the typical temperature evolution (see Figure 3.6) where the exchanging lines undergo broadening and coalescence [5]. Finally, fast exchange results in observation of a single resonance detected at an averaged frequency: $\nu = (\nu^A + \nu^X)/2$. In this situation the spectral parameters of individual states as well as their relaxation times, $T_{1,2}(A)$ and $T_{1,2}(X)$ are completely averaged.

The spin states A and X can have different populations and belong to the same molecule (intramolecular exchange) or different molecules (intermolecular exchange). Also a natural relaxation time can be significantly shorter in one of the spin states. Then it is obvious that the positional exchange will more strongly affect the other state with longer $T_{1,2}$ times. The effect is particularly remarkable when one of the states is weakly populated, but paramagnetic.

A positional exchange is a *slow* when $\nu_E < \nu^A - \nu^X$. Under these conditions, the A and X lines are well separated and remain narrow (see Figure 3.6). However, in spite of the good separation of the resonances, their relaxation times can be *partially* averaged. Owing to this effect, free induction decays, collected by the inversion recovery experiments, will depend on $T_1(A)$ and $T_1(X)$ times and in addition, on lifetimes of spins in the both states: τ^A and τ^X. For these reasons, T_1 calculations from the inversion recovery data will require separations of two superimposed exponentials. In turn, standard treatments of these data in a monoexponential approximation will lead to

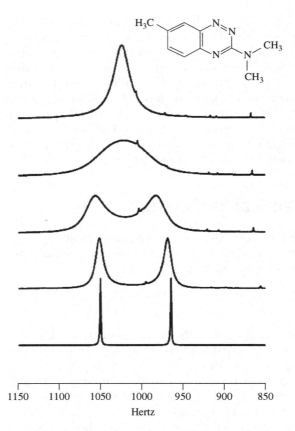

Figure 3.6 Temperature evolution of the NCH$_3$ lineshapes in the variable-temperature ^{1}H NMR spectra due to rotation around the C$-$N bond on the NMR timescale. (Reproduced from A. D. Bain. *Progress in Nuclear Magnetic Resonance Spectroscopy* 2003; **43**: 63, with permission from Elsevier)

effective magnitudes T_1(eff). Sometimes natural T_1 times can be determined from the T_1(eff) values with the help of approximate formulas [6]. Note, however, that in this case exchange rates $1/\tau^A$ and $1/\tau^X$ should be obtained independently. In the absence of such information, slow positional exchanges, distorting the natural T_1 values, are not desirable.

Slow positional exchanges can be identified spectroscopically by *saturation transfer experiments* in ^{1}H NMR spectra. The experiments involve a double-resonance technique and consist of observation of A nuclei during irradiation of X nuclei. In the presence of a saturating radiofrequency field, fast transitions of X nuclei between the Zeeman levels equalize their populations and the X resonance is *saturated*. Owing to a slow positional exchange, the saturating effect can be transferred from X to A. The latter is accompanied

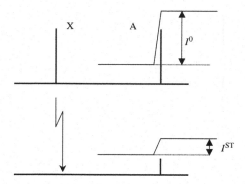

Figure 3.7 Schematic Presentation of a saturation transfer experiment on A and X nuclei coupled to a slow positional exchange: I^0 is an equilibrium integral intensity of the A resonance; I^{ST} an integral intensity after irradiation of the X resonance

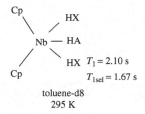

$T_1 = 2.10$ s
$T_{1sel} = 1.67$ s

toluene-d8
295 K

Figure 3.8 The selective (T_{1sel}) and non-selective (T_1) ^{1}H spin–lattice relaxation times measured for the H(X) ligand of the Nb trihydride complex in the presence of a slow H(X)/H(A) exchange (400 MHz, toluene-d$_8$)

by decreasing the integral intensity of the A resonance I^{ST}, as shown in Figure 3.7.

The spectral behavior of the Nb trihydride complex (Figure 3.8) in solution illustrates the importance of the saturation transfer studies. The room-temperature ^{1}H NMR spectrum of the trihydride shows two well-separated hydride resonances, H(X) and H(A) (Figure 3.9). According to the molecular structure of the complex, the H(A) hydride atom is adjacent to two equivalent H(X) ligands. As we will show in Chapter 4, two dipole–dipole H(A)–H(X) contacts should strongly reduce the T_1(HA) time with respect to T_1(HX). In contrast, the measured T_1(HA) and T_1(HX) values are very similar. A partial T_1 averaging (due to an H(A)/H(X) exchange) could cause this effect. However, the NMR spectrum in Figure 3.9 does not reveal the presence of the exchange. In contrast, the saturation transfer experiments, performed for the H^X/H^A pair, have shown directly the presence of a slow hydride/hydride exchange. It is obvious that, under these conditions, the T_1(HA) and T_1(HX) values cannot be used as individual natural characteristics of the hydride resonances and the temperature of T_1 measurements should be decreased.

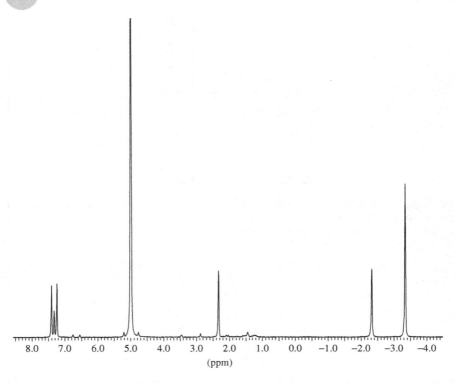

Figure 3.9 The room-temperature ^1H NMR spectrum of complex Cp_2NbH_3 (Figure 3.8) in toluene-d_8 at 400 MHz. The signals in the region of 7.5 and 2.5 ppm belong to the solvent

For other nuclei than protons, the saturation transfer technique is not applied. In such cases, the presence of slow positional exchanges can be established with the help of the two-dimensional NOE (NOESY) NMR experiments, known as two-dimensional exchange spectroscopy (EXSY) [7].

Slow positional exchanges can be successfully revealed by ^1H T_{1sel}/^1H T_1 measurements [8]. Theoretically, the dipolar T_{1sel} times are usually longer than the nonselective T_1 times. The hydride resonances of the Nb complex in Figure 3.8 demonstrate an opposite tendency and $T_{1sel}(HX) < T_1(HX)$. The selective inverting 180° pulse, operating at the H(X) resonance frequency, results in different spin temperatures of the H(A) and H(X) states. Then, an additional relaxation rate for the H(X) resonance, appearing in the selective experiments, is explained by a slow exchange between the H(X) spins and their 'colder' neighbor.

Finally, we illustrate an 'abnormal' relaxation behavior of NMR signals caused by a chemical exchange involving paramagnetic species. The diamagnetic 18-e complex Cp*FeH(dppe) is protonated with weak acids (for

Table 3.1 ^{1}H T_1 relaxation data in a CD_2Cl_2 solution of Cp*FeH(dppe) in the presence of a three-fold excess of 2,2,2-trifluoroethanol (400 MHz)

T (K)	$T_1(Fe(H_2))$(ms)	$T_1(FeH)$(ms)
190	24	25
200	15	12
220	11	3.9
240	9.6	0.67
250	11	

example, 2,2,2-trifluoroethanol) in CD_2Cl_2 solutions to give the diamagnetic dihydrogen complex [9]:

$$Cp^*FeH(dppe) + H^+ \longleftrightarrow Cp^*Fe(H_2)(dppe)^+ \qquad (3.2)$$

Both hydride resonances (FeH and Fe(H$_2$)), well observed in the ^{1}H NMR spectra, provide the accurate T_1 time measurements, shown in Table 3.1. Theoretically, dipolar proton–proton coupling and the chemical shift anisotropy mechanism (see Chapter 5) govern nuclear relaxation of the hydride resonances. In accordance with the theory of dipole–dipole nuclear relaxation (see Chapter 4), the ^{1}H $T_1(FeH_2)$ time, measured for the dihydrogen ligand, is short, due to strong H–H dipolar coupling. The time increases on cooling and goes through a minimum of 9.6 ms at 240 K again in accordance with the theory. In contrast, the relaxation behavior of the hydride resonance in the initial complex is abnormal. Actually in the absence of strong dipole–dipole proton–proton interactions, the T_1(Fe-H) times are unusually short particularly in the region of 240 K (0.7 ms). For non-quadrupolar nuclei such magnitudes are direct evidence for the presence of paramagnetic species involved in an exchange with complex Cp*FeH(dppe). In fact, the source of the paramagnetic relaxation is the 17-e paramagnetic complex [Cp*FeH(dppe)]$^+$ X$^-$ which is unobservable by the ^{1}H NMR spectra because of a small concentration and the paramagnetic nature. The paramagnetic complex participates in the degenerate exchange:

$$18e\text{-}Cp^*FeH(dppe) + 17e\text{-}[Cp^*FeH(dppe)]^+X^-$$

$$\Longleftrightarrow 17e\text{-}[Cp^*FeH(dppe)]^+X^- + 18e\text{-}Cp^*FeH(dppe) \qquad (3.3)$$

affecting the ^{1}H T_1 times measured for the diamagnetic complex. On heating, the rate of the exchange increases and the observed effects become more pronounced. It is obvious that, under these circumstances, the measured T_1 times cannot be used for structural characterizations of the investigated compounds.

Bibliography

1. V. A. Daragan, M. A. Loczewiak, K. H. Mayo. *Biochemistry* 1993; **32**: 10580.

2. S. Schaublin, A. Hohener, R. R. Ernst. *Journal of Magnetic Resonance* 1974; **13**: 196.

3. V. I. Bakhmutov, E. V. Vorontsov, G. Boni, C. Moise. *Inorganic Chemistry* 1997; **36**: 4055.

4. D. M. Grant, C. L. Mayne, F. Liu, T. X. Xiang. *Chemical Reviews* 1991; **91**: 1591.

5. A. D. Bain. *Progress in Nuclear Magnetic Resonance Spectroscopy* 2003; **43**: 63.

6. H. Strehlow, J. Frahm. *Berichte den Bunsengesellschaft für Physitalische Chemie* 1975; **79**: 57.

7. J. K. M. Sanders, B. K. Hunter. *Modern NMR Spectroscopy. A Guide for Chemists.* Oxford, New York, Toronto: Oxford University Press, 1994.

8. V. I. Bakhmutov, E. V. Vorontsov, G. I. Nikonov, D. A. Lemenovskii. *Inorganic Chemistry* 1998; **37**: 279.

9. N. V. Belkova, P. O. Revin, L. Epstein, E. V. Vorontsov, V. I. Bakhmutov, E. S. Shubina, E. Collange, R. Poli. *Journal of the American Chemical Society* 2003; **125**: 11106.

4 NMR Relaxation by Dipole–Dipole and Quadrupole Interactions

The fluctuating magnetic fields, causing nuclear relaxation, are created in a sample by dipole–dipole, quadrupole, spin-rotation, scalar and chemical shift anisotropy interactions. For many nuclei, different relaxation mechanisms can operate simultaneously. In such cases, each mechanism provides the corresponding contribution to a total relaxation rate. In addition, two 'independent' mechanisms can interfere to give the *cross-relaxation* contributions.

In spite of the variety of relaxation mechanisms and the different origin of nuclear coupling, relaxation rates $1/T_1$ (and also $1/T_2$ or $1/T_{1\rho}$) can be expressed via the general formula:

$$1/T_1 = C\,J(\omega_0, \tau_C) \tag{4.1}$$

Practical NMR Relaxation for Chemists   Vladimir I. Bakhmutov
© 2004 John Wiley & Sons, Ltd   ISBNs: 0-470-09445-1 (HB); 0-470-09446-X (PB)

where C is the strength of nuclear coupling. The function $J(\omega_0, \tau_C)$, called *the spectral density function*, shows how the influence of nuclear coupling C on the relaxation rate changes when frequencies of the field fluctuations move off the Larmor frequency, $\omega_0(\omega_0 = 2\pi\nu_0)$. In other words, C connects with molecular structure while $J(\omega_0, \tau_C)$ depends on molecular mobility.

4.1   Intramolecular Dipole–Dipole Relaxation: Homo- and Heteronuclear Dipolar Coupling and the Spectral Density Function

Interactions of nuclei of one sort with nuclei of other sort are called *heteronuclear* dipolar coupling. *Homonuclear* coupling corresponds to interactions of nuclei of the same sort. If the magnetic dipole of one nucleus (see Equation 1.5), for example, a proton, can interact with the magnetic dipole of nucleus B, other than a proton, then the strength of this heteronuclear dipolar coupling, DC, measured in Hz, is:

$$DC_{H-B} = (4/30)(\mu_0/4\pi)^2\, r(H-B)^{-6}\, \gamma_H{}^2\gamma_B{}^2\, \hbar^2 I_B(I_B + 1) \qquad (4.2)$$

where γ_H and γ_B are the nuclear magnetogyric ratios of ^1H and B nuclei, respectively, I_B is the spin of nucleus, B μ_0 is the permeability of vacuum and $r(H-B)$ is the internuclear distance [1]. Note that the equation is written to account for 100% natural abundance of nucleus B. When the natural abundance of target nuclei less than 100%, (for example, ^{11}B), the equation can be easily modified.

Equation (4.2) shows that the strength of dipolar nuclear coupling is proportional to the inverse sixth power of the internuclear distance. It is easy to demonstrate that the dipolar coupling reduces by *244 times* with increasing the internuclear distance from 1 to 2.5 Å. It becomes obvious that dipolar coupling is significantly more effective for nuclei, located in the same molecule. Thus, in this case the DC constant is directly related to molecular structure.

When two coupled nuclei are identical, for example, protons, the strength of the homonuclear dipolar coupling, DC_{H-H}, is written as:

$$DC_{H-H} = 0.3\,(\mu_0/4\pi)^2\, \gamma_H{}^4\, \hbar^2\, r(H-H)^{-6} \qquad (4.3)$$

The DC constants in Equations (4.2) and (4.3) change proportionally to γ^2. This fact and the magnetic properties in Table 1.1 allow us to predict the role of dipole–dipole interactions in relaxation of different nuclei. It is obvious that the strength of dipolar coupling is highest for a proton pair or a pair where one of the nuclei is a proton. Significant dipole–dipole interactions can be also expected for ^{19}F nuclei and nuclei in their environment.

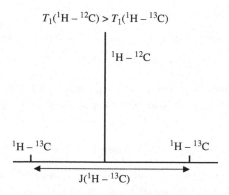

$$T_1(^1H - {}^{12}C) > T_1(^1H - {}^{13}C)$$

Figure 4.1 ^1H NMR spectrum of a C–H fragment (schematic). Intensity of each satellite line is 0.5% due to the natural abundance of the ^{13}C isotope

Effects of dipolar interactions on relaxation of proton/carbon pairs depend on target nuclei. In fact, because of spin–spin scalar coupling and the low natural abundance of ^{13}C isotopes (~1%), the ^1H NMR spectrum of a C–H fragment (Figure 4.1) exhibits a central line and two low–intensity satellite lines. Theoretically, the ^1H T_1 time of these satellite lines should be shorter than that of the central resonance, owing to the presence of the ^1H–^{13}C dipolar coupling. However, for obvious reasons, the objects of ^1H relaxation experiments are the central resonances. They correspond to protons attached to the magnetically inactive isotope ^{12}C and show no proton–carbon dipolar contributions. In contrast, ^{13}C isotopes are always attached to ^1H and therefore C–H dipole–dipole interactions make one of the major contributions to ^{13}C T_1 rates. Finally, in most cases, dipolar coupling will be effective in nuclear relaxation if DC values are between 10^4 and 10^5 Hz.

The properties of ^1H and ^2H isotopes predict weakening dipolar coupling on displacement of protons by deuterons. This is why isotopic displacement is often applied in relaxation experiments to reduce or evaluate contributions of the dipole–dipole relaxation [2]. For the same reason, application of deuterated solvents minimizes the solvent effects in ^1H NMR relaxation experiments. For example, dipole–dipole interactions between the hydride ligand and deuterons of a solvent contribute only 3% to the ^1H T_1 relaxation rate of $HRe(CO)_5$ in C_6D_6 [3].

Figure 1.13 explains the appearance of field fluctuations due to reorientations of dipolar vectors with respect to the direction of the external magnetic field. When these reorientations (or in other words, molecular motions) have large amplitudes and *random* character, then the spectral density function, $J(\omega_0, \tau_C)$, takes the Bloembergen–Purcell–Pound form [1] and the rate of spin–lattice relaxation $(1/T_1(H...B))$ by the heteronuclear dipole–dipole interactions is:

$$1/T_1(H\ldots B) = DC_{H-B}\{3\tau_c/(1+\omega_H^2\tau_c^2) + 6\tau_c/[1+(\omega_H+\omega_B)^2\tau_c^2)$$
$$+ \tau_c/(1+(\omega_H-\omega_B)^2\tau_c^2]\} \quad (4.4)$$

where ω_H and ω_B are the resonance frequencies of H and B nuclei, respectively. By analogy, the rate of the proton–proton dipole–dipole relaxation is expressed as:

$$1/T_1(H\ldots H) = DC_{H-H}[\tau_c/(1+\omega_H^2\tau_c^2) + 4\tau_c/(1+4\omega_H^2\tau_c^2)] \quad (4.5)$$

Since the molecular motion correlation time, τ_c, depends on the temperature ($\tau_C = \tau_0 \exp(E_a/RT)$), the plots of Equations (4.4) and (4.5) in semilogarithmic coordinates are symmetrical V-shaped curves with minima (Figure 4.2). It is easy to show that in the case of ^1H relaxation, the $J(\omega_0, \tau_C)$ term takes a minimal value at:

$$\tau_C = 0.62/\omega_H \quad (4.6)$$

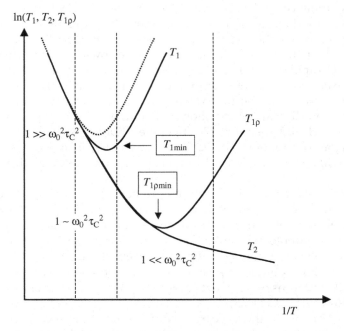

Figure 4.2 Temperature dependences of the dipolar T_1, T_2 and $T_{1\rho}$ relaxation times in semilogarithmic coordinates at the Larmor frequency ω_0. The dashed T_1 curve corresponds to a higher Larmor frequency than ω_0. The regions with $1 \gg \omega_0^2\tau_C^2$, $1 \sim \omega_0^2\tau_C^2$ and $1 \ll \omega_0^2\tau_C^2$ correspond to fast, intermediate and slow molecular motions on the frequency scale of NMR

According to this relationship, the correlation time is proportional to the $1/\omega_H$ and thus the temperature, corresponding to the $^1H\ T_{1\,min}$ value, increases with increasing the strength of the external magnetic field. In addition, location of T_1 minima on the variable-temperature relaxation curves directly depends on molecular mobility. Therefore, in solutions, where molecular motions are fast, $^1H\ T_{1\,min}$ times are observed in low-temperature experiments only on relatively bulky molecular systems or highly viscous solvents, for example, glycerin [4]. The power of an effective radiofrequency field, operating in the $T_{1\rho}$ spin-locking experiments, is significantly less than that of the external magnetic field. For this reason, in accordance with Equation (4.6), $T_{1\rho}$ minima should be observed at significantly lower temperatures (Figure 4.2) with respect to $T_{1\,min}$. In practice, $T_{1\rho}$ minima are reached relatively easy in solids where molecular motions are slow. In solutions, these experiments can be successful only in the case of very big molecules.

The left and right wings of the T_1 plots in Figure 4.2 correspond to $\omega_H^2\tau_C^2 \ll 1$ (known as the extreme narrowing condition or fast motional regime) and $\omega_H^2\tau_C^2 \gg 1$ (slow molecular motions), respectively. Under these conditions, Equation (4.5) converts to the equations:

$$1/T_1(\mathrm{H}\ldots\mathrm{H}) = (3/2)(\mu_0/4\pi)^2\ \gamma_H^4\ \hbar^2\ r(\mathrm{H-H})^{-6}\ \tau_c \qquad (4.7)$$

$$1/T_1(\mathrm{H}\ldots\mathrm{H}) = (3/5)(\mu_0/4\pi)^2\ \gamma_H^4\ \hbar^2\ r(\mathrm{H-H})^{-6}(1/\omega_H^2\tau_C) \qquad (4.8)$$

In turn, for minimal $^1H\ T_1$ values one can write:

$$1/T_{1\,min}(\mathrm{H}\ldots\mathrm{H}) = 0.427(\mu_0/4\pi)^2\ \gamma_H^4\ \hbar^2\ r(\mathrm{H-H})^{-6}/\omega_H \qquad (4.9)$$

As seen from Figure 4.2, T_1 times are independent of the working frequency of NMR spectrometers in the extreme narrowing region, while they increase proportionally to ω_0 at low temperatures. Equations (4.7) and (4.8) show that in fast and slow motional regions, the plots of $\ln T_1$ versus $1/T$ are linear and their slopes correspond to activation energies of molecular motions.

Since the mechanisms of spin–lattice and spin–spin nuclear relaxation are different in principle, their temperature dependences are also different. In the case of heteronuclear spin–spin relaxation, the spectral density function, $J(\omega_0, \tau_C)$, takes the form:

$$J(\omega_0, \tau_C) = [4\tau_c + 3\tau_c/(1 + \omega_H^2\tau_c^2) + 6\tau_c/(1 + \omega_B^2\tau_c^2)$$
$$+ \tau_c/(1 + (\omega_H - \omega_B)^2\tau_c^2) + 6\tau_c/(1 + (\omega_H + \omega_B)^2\tau_c^2)] \qquad (4.10)$$

Combining the $J(\omega_0, \tau_C)$ functions in Equations (4.4) and (4.10) shows that at $\omega_H^2\tau_C^2 \ll 1$ the T_2 and T_1 times are identical (Figure 4.2). However, on cooling, the T_2 time reduces monotonically. This effect leads to the well-known strong broadenings of NMR signals at low temperatures.

Finally, all these equations are written in terms of a single dipole–dipole contact. The relaxation rates are additive and therefore, in the presence of a larger number of such contacts, all the dipolar contributions should be summarized:

$$1/T_1^{OBS} = \sum_n 1/T_1 \qquad (4.11)$$

where n is the number of corresponding dipole–dipole contacts.

4.2 How to Reveal the Presence of the Dipolar Mechanism

Dipole–dipole NMR relaxation has two important phenomenological features: (i) T_1 times do not depend on the Larmor frequencies and increase in high-temperature regions; (ii) minimal T_1 values are proportionally to the Larmor frequencies. As we will show below, the spin-rotation and chemical shift anisotropy mechanisms demonstrate another behavior. Therefore, the above-mentioned features can be successfully used to identify the dipole–dipole relaxation. Note, however, that in the presence of different relaxation mechanisms direct evidence for dipole–dipole interactions can be obtained by NOE experiments and measurements of the selective relaxation times.

4.2.1 NOE as a Test for Dipole–Dipole Nuclear Relaxation

By definition, the nuclear Overhauser effect (NOE) is observed for a pair of A/X nuclei if they are coupled by dipole–dipole interactions [4]. The effect consists of perturbation of an equilibrium magnetization, measured for A nuclei as the equilibrium integral intensity of the NMR signal, when X nuclei are irradiated by a radiofrequency field (Figure 4.3). At positive magnetogyric ratios of A and X nuclei, the effect is responsible for an *enhancement* of the A integral intensity.

Since the correlation time of molecular tumbling depends on the temperature, the NOE is also temperature dependent. Figure 4.4 illustrates this dependence for a pair of protons as the plot of the NOE enhancement versus $\omega_0\tau_C$. Increasing the τ_C values (corresponding to decreasing the temperature) leads to reducing the positive NOE enhancement to zero and then, to increasingly negative NOE values. It is known that small molecules are reoriented in nonviscous solutions at the τ_C values of $10^{-10}-10^{-11}$ s. It is obvious that the NOE enhancements for these molecules should be positive at moderate temperatures. In contrast, big molecules, such as proteins, tumble at τ_C values of 10^{-7} s. In such cases, the NOE is negative.

The NOE magnitudes can be negative if one of the coupled nuclei has a negative magnetogyric ratio, for example, as in the $\{^1H\}/^{15}N$ or $\{^1H\}/^{29}Si$

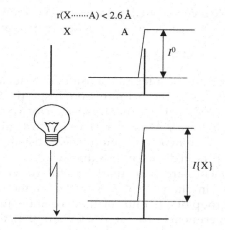

Figure 4.3 NOE enhancement observed for a pair of A/X protons coupled by dipolar interactions and separated by a relatively short internuclear distance where X is the irradiated nucleus and A is the detected one

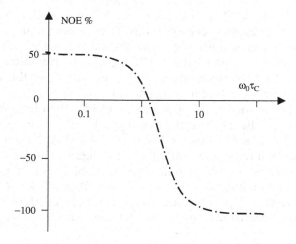

Figure 4.4 Nuclear Overhauser effect, observed in a pair of protons, as a function of the molecular motion correlation times τ_C. Increasing the $\omega_0\tau_C$ values corresponds to decreasing the temperature

pairs. In addition a system consisting of three protons:

$$H(A)\ldots H(B)\ldots H(C) \tag{4.12}$$

also shows an unusual NOE behavior. In fact, the H(A) irradiation results in the *positive* and *negative* NOE enhancement observed for H(B) and H(C) nuclei, respectively.

If the dipole–dipole mechanism dominates in relaxation of A and X nuclei, the NOE observed for X nuclei can be expressed by the ratio:

$$M^X\{A\}/M^X_0 = 1 + \gamma_A/2\gamma_X \qquad (4.13)$$

where $M^X\{A\}$ and M^X_0 are the integral intensities of X nuclei measured in the presence and the absence of the A irradiation, respectively. Equation (4.13) shows that a maximal NOE enhancement, expected for a pair of protons (i.e. at $\gamma_A = \gamma_X = \gamma_H$) is equal to 50%. However, since the strength of dipole–dipole interactions reduces with increasing internuclear distance, the NOE can be observed at proton–proton distances less than 2.6 Å.

The NOE experiments are particularly useful when several relaxation mechanisms operate simultaneously. If X nuclei interact with A nuclei and in addition, with a group of other nuclei, then a combination of T_1 measurements and NOE experiments provide to evaluate the A–X dipole–dipole contribution according to:

$$M^X\{A\}/M^X_0 = 1 + (\gamma_A/2\gamma_X) \times (T_1DD(X\text{-}A))^{-1}/T_1(TOT)^{-1} \qquad (4.14)$$

This situation often takes place in ^{13}C NMR [6] where the NOE measured in ^{13}C and ^{13}C{^{1}H} NMR experiments, leads to a reliable evaluation of the dipolar C–H relaxation contributions. Finally note that in practice the NOE experiments can be performed with the help of one-dimensional (with selective saturation of one of the NMR signals) or two-dimensional (for example, NOESY) NMR techniques [7]. In the last case, the cross-peaks, responsible for NOEs (see Figure 1.4), should be analyzed.

As an example of the NOE application, consider the niobium trihydride (Figure 4.5) where the H(X) and H(A) ligands are separated by 1.98 Å. Theoretically, the ^{1}H spin–lattice relaxation of the H(A) ligand is governed by dipole–dipole interactions with the H(X) ligands and protons in the Cp rings. In addition, dipole–dipole proton–niobium interactions are also effective. A Cp irradiation does not affect intensities of the H(A) and H(X) resonances and thus Cp contributions to T_1(HA) and T_1(HX) times are negligible. In contrast, irradiating the H(X) nuclei leads to a very pronounced

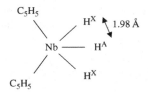

Figure 4.5 The niobium trihydride complex (schematically), containing the lateral (H^X) and central (H^A) hydride ligands. The ligands are spatially separated by 1.98 Å and show ^{1}H resonances with the chemical shift difference of ∼1 ppm (toluene-d$_8$). Both circumstances provide good conditions for quantitative NOE measurements [8]

NOE observed for the H(A) resonance (Figure 4.6) [8]. The quantitative NOE measurements, in combination with relaxation experiments, give finally a 50% dipole–dipole H(A)–H(X) contribution to the H(A) relaxation rate.

The ^{31}P and $^{31}P\{^1H\}$ NMR spectra, recorded for the organophosphorus compounds in Figure 4.7, reveal the small (but nonzero) NOE enhancements [9]. These values and the relaxation measurements result in calculations of the dipole–dipole ^{31}P T_1 times (Table 4.1). The ^1H–^{31}P coupling is too weak to provide the effective relaxation mechanism for ^{31}P nuclei.

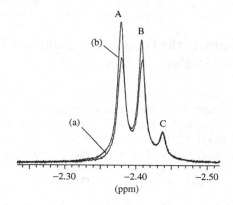

Figure 4.6 H^A resonances in the ^1H NMR spectrum of the trihydride Nb complex (see Figure 4.5) and its deuterated derivatives without (a) and with (b) irradiation of the H^X signal in toluene-d_8, at 220 K and 400 MHz; A = $Cp_2NbH^X_2H^A$, B = $Cp_2NbH^XD^XH^A$, C = $Cp_2NbD^X_2H^A$. (Reproduced with permission from V. I. Bakhmutov, E. V. Vorontsov, G. I. Nikonov, D. A. Lemenovskii. *Inorganic Chemistry* 1998; **37**: 279. © 1998 American Chemical Society)

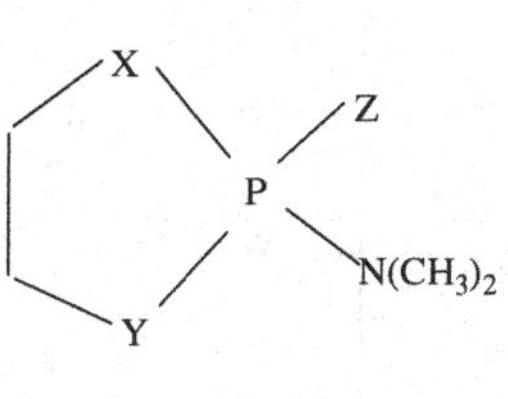

1 X = Y = O, Z = Lone pair

2 X = Y = NCH₃, Z = Lone pair

3 X = Y = NCH₃, Z = BH₃

4 X = O, Y = NCH₃, Z = Lone pair

5 X = O, Y = NCH₃, Z = BH₃

Figure 4.7 Dimethylaminephospholanes and their borane adducts investigated by ^{31}P T_1, T_2 and NOE measurements (Table 4.1)

Table 4.1 Total spin–lattice (T_1), spin–spin (T_2), dipolar (T_1DD) and scalar (T_2SC) ^{31}P relaxation times in compounds 1–5 (Figure 4.7) measured in $CDCl_3$ at room temperature

Compound	T_1 (s)	T_2 (s)	T_2SC (s)	T_1DD (s)
1	10.2 ± 0.1	0.043 ± 0.002	0.043 ± 0.002	1792 ± 100
2	11.8 ± 0.1	0.01 ± 0.0005	0.01 ± 0.0005	242 ± 12
3	11.5 ± 0.3	0.063 ± 0.003	0.063 ± 0.003	471 ± 28
4	10.7 ± 0.3	0.01 ± 0.0005	0.01 ± 0.0005	62.6 ± 4
5	11.06 ± 0.04	0.038 ± 0.001	0.038 ± 0.001	45.3 ± 4

4.2.2 Evaluations of the Dipolar Contributions from Selective and Nonselective T_1 Times

Consider the spin system, consisting of a pair of the A and B protons. The protons are spatially approximated and coupled. Figure 4.8 illustrates the energy states, that come to be combined, and are responsible for a cross-relaxation phenomenon [10]. When an inverting 180° pulse in the inversion recovery experiment excites the B nucleus selectively, then its T_1 time is:

$$1/T_{1sel}(B) = 2W_B + W_2 + W_0 \tag{4.15}$$

where W_B, W_2 and W_0 are probabilities of the corresponding nuclear transitions. If this first 180° pulse is not selective, then:

$$1/T_1(B) = 2W_B + 2W_2 \tag{4.16}$$

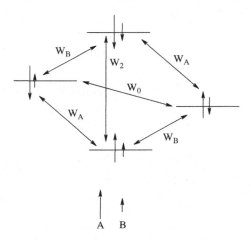

Figure 4.8 Scheme of energy levels for a two-spin AB system where W is the probability of the corresponding nuclear transitions

Thus, theoretically, the T_{1sel}/T_1 ratio is equal to 1.5. It is obvious that this value, observed experimentally, will be a good test for a 100% dipolar relaxation of the target nuclei. Smaller T_{1sel}/T_1 values will reveal the presence of additional relaxation mechanisms.

4.3 Intermolecular Dipole – Dipole Interactions

Local magnetic fields, fluctuating in the lattice due to translational molecular motions, lead to relaxation of spins, located in different molecules (intermolecular dipole–dipole interactions). If molecules are spherical and translational molecular motions are fast, then the relaxation rate of I spins, $1/T_1(I)$, is:

$$1/T_1(I) = (\mu_0/4\pi)^2 \, (8/45) \, N_S \, \gamma_I^2 \, \gamma_S^2 \, \hbar^2 \, S(S+1)/D_{IS} \, r(I-S) \qquad (4.17)$$

where N_S is the concentration of S spins, $r(I-S)$ is the closest internuclear distance and D_{IS} is the translation self-diffusion constant. By definition, this type of nuclear relaxation characterizes molecular mobility or intermolecular interactions. In the context of structural applications, this mechanism is undesirable. In fact, molecular collisions of solute–solute or solute–solvent will shorten relaxation times. In order to reduce these effects, concentrations of investigated solutions should be decreased.

4.4 Electric Field Gradients at Quadrupolar Nuclei

The quadrupole mechanism dominates spin–lattice and spin–spin relaxation of nuclei, having spin numbers $> 1/2$ [1]. Distributions of charges at such nuclei are nonspherical (Figure 4.9) and responsible for appearance of nuclear quadrupole moments Q. In this situation, the spins of quadrupole nuclei can interact with not only the external and local magnetic fields, but also with any *electric field gradients* at these nuclei. Most of quadrupolar nuclei (for example, ^{81}Br, ^{127}I or ^{35}Cl) have large quadrupole moments. These nuclei become objects of nuclear quadrupole resonance (NQR) studies. By analogy with NMR, NQR experiments are based on quantized energy levels, corresponding to different orientations of quadrupole moments Q with respect to the electric field gradients (EFG).

The EFG (for example, eq_{ZZ}, directed along a chemical bond) is expressed via the electrostatic potential V as:

$$eq_{ZZ} = \partial^2 V / \partial_Z^2 \qquad (4.18)$$

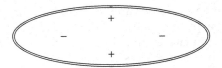

Figure 4.9 Schematic presentation of a non-spherical charge distribution responsible for appearance of the nuclear quadrupole moment

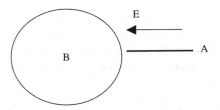

Figure 4.10 A non-homogeneous electric field, E (schematically), changing along the chemical bonds, A–B, and characterized by the eq_{ZZ} values

Thus, the EFG characterizes a non-homogeneous electric field E, changing along this chemical bond (see atoms A and B in Figure 4.10). The electrostatic potential V is a scalar magnitude. In contrast, the electric field gradient is expressed mathematically as a tensor:

$$\begin{vmatrix} eq_{XX} & 0 & 0 \\ 0 & eq_{YY} & 0 \\ 0 & 0 & eq_{ZZ} \end{vmatrix} \quad (4.19)$$

where the off-diagonal elements are equal to zero. By convention, the largest element of the EFG tensor, eq_{ZZ}, is always oriented along the A–B bond (the Z-axis). Finally, according to the theory of electrostatic interactions, the trace of the tensor $(q_{ZZ} + q_{YY} + q_{XX})$ should be equal to zero.

4.5 Nuclear Quadrupole Coupling Constant as Measure of the Electric Field Gradient

The energy of interactions between the nuclear quadrupole moments Q and the EFG is expressed through the nuclear quadrupole coupling constant (NQCC) [11]:

$$NQCC = e^2 q_{zz} Q/h \quad (4.20)$$

where eq_{ZZ} is the principal component of the EFG tensor, e is the elementary charge ($1.6021773 \times 10^{-19}$ C) and NQCC is measured in frequency units and is of the order of $10^6 - 10^9$ Hz. Thus, the NQCC is a measure of the 'size' of the

EFG. A spatial extension or 'shape' of the EFG is defined via the asymmetry parameter η:

$$\eta = |eq_{XX} - eq_{YY}|/eq_{ZZ} \tag{4.21}$$

When the XX and YY components are equal (i.e. $eq_{XX} = eq_{YY}$ and $\eta = 0$), then the electric field gradient is axially symmetric.

Shapes and sizes of the electric field gradients at nuclei depend strongly on the symmetry of charge distributions around these nuclei [12, 13]. Figure 4.11 helps to rationalize large variations in NQCC values, shown in Table 4.2. For example, a symmetric charge distribution at the ^{14}N nuclei in $NH_4^+Cl^-$ reduces the NQCC value to 0.016 MHz compared with 0.9 or even 3.98 MHz in such compounds as $EtONO_2$ or $MeNH_2$, respectively. The NQCC value is close to zero at the deuterium ligand in solid hydride PdD or at ^{11}B nuclei in the BH_4^- ions again due to the fully symmetric charge distributions.

On a semiquantitative level, the electric field gradient at any quadrupolar nucleus can be expressed as the sum of nuclear and electronic terms:

$$eq_{ZZ} = +\sum_{n} K_n(3z_n{}^2 - r_n{}^2)/r_n{}^5 - e\langle\psi^*| \sum_{i}(3z_i{}^2 - r_i{}^2)/r_i{}^5|\psi\rangle \tag{4.22}$$

where K and e are charges on the neighboring nuclei and electrons, respectively, and r_n and r_i are the corresponding distances. Thus, the electric field gradient at the A nucleus in the A—B bond (Figure 4.10) depends on the B charge as well as the A—B bond length. In other words, the NQCCs characterize the A—B bonding modes.

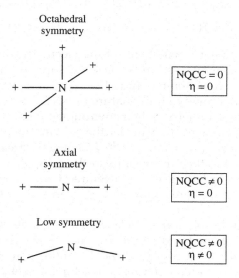

Figure 4.11 NQCC and η variations as a function of the charge distribution symmetry around the registered nuclei N

Table 4.2 NQCC variations for ^{17}O, ^{14}N, ^2H and ^{11}B

Compound	Nucleus	NQCC (MHz)
O_2	^{17}O	−8.42
CO	^{17}O	4.43
$MoO_4{}^{2-}$	^{17}O	0.7
$NaNO_3$	^{14}N	0.745
CH_3CN	^{14}N	4.00
$MeNH_2$	^{14}N	3.98
$EtONO_2$	^{14}N	0.9
NH_4Cl	^{14}N	0.016
solid PdD	^2H	0.000
$BD_3\,NH_3$	^2H	0.105
$BH_4{}^-$	^{11}B	0.0

4.6 Quadrupole Relaxation

Spins I of quadrupolar nuclei, placed in an external magnetic field, inter-act with the electric field gradients, eq_{ZZ}, oriented usually along chemical bonds. Random rotational reorientations of these bonds lead to the appear-ance of fluctuating magnetic fields. Under these conditions, the spin–lattice relaxation time, $T_1(Q)$, is:

$$1/T_1(Q) = (3/50)\pi^2(2I+3)(I^2(2I-1))^{-1}(e^2 q_{zz}Q/h)^2(1+\eta^2/3)$$
$$\times (\tau_c/(1+\omega_Q{}^2\tau_c{}^2) + 4\tau_c/(1+4\omega_Q{}^2\tau_c{}^2)) \quad (4.23)$$

where η is the asymmetry parameter of the electric field gradient and I is the nuclear spin. Table 4.3 lists $T_1(Q)$ times, measured for some quadrupolar nuclei in solutions at room temperature. The $T_1(Q)$ times vary over very large ranges and depend strongly on the nature of nuclei (values of quadrupole moments Q) and the symmetry of their environments. For example, in spite of the big quadrupole moment of ^{14}N nuclei, their environment in the compound $Me_4{}^{14}NBr$ is fully symmetric and the ^{14}N T_1 time is greatly increased (up to 10^4 ms). In contrast, the NH_3 groups in the complex $(Co(NH_3)_6)(ClO_4)_3$ show a very short ^{14}N T_1 time (0.3 ms). Similar tendencies are observed for ^{127}I and ^{35}Cl nuclei (compare compounds SnI_4 and $IF_6{}^+$ or HCl and $ClO_4{}^-$, respectively).

As in the case of dipole–dipole relaxation, temperature dependences of $T_1(Q)$ times are V-shaped and symmetric in semilogarithmic coordinates. When $\tau_C = 0.62/\omega_Q$, they also go through minima. In the region of fast molecular motions (high-temperature zones) Equation (4.23) converts to:

$$1/T_1(Q) = 0.3\pi^2(2I+3)[I^2(2I-1)]^{-1}(e^2 q_{zz}Q/h)^2(1+\eta^2/3)\tau_c \quad (4.24)$$

Table 4.3 Quadrupole nuclear moments (Q) and the room-temperature $T_1(Q)$ times of some quadrupolar nuclei measured in solution

Compound	Nucleus (Q) ($10^{28} Q/m^2$)		$T_1(Q)$ (ms)
HBr	^{81}Br	(0.31)	14.2×10^{-3}
PBr_3	^{81}Br		0.3×10^{-3}
SnI_4	^{127}I	(−0.79)	0.15×10^{-3}
IF_6^+	^{127}I		1
$CFCl_3$	^{35}Cl	(−0.10)	38.3×10^{-3}
HCl	^{35}Cl		0.8
ClO_4^-	^{35}Cl		270
MeNC	^{14}N	(1×10^{-2})	1220
Me_4NBr	^{14}N		10^4
$(Co(NH_3)_6)(ClO_4)_3$	^{14}N		0.3
closo-2,4-$C_2B_5H_7$	^{11}B	(4.1×10^{-2})	38.6–68.3
nido-$B_{10}H_{14}$	^{11}B		4.8–17.3
$C_6D_5CD_3$	^2H	(2.8×10^{-3})	5×10^3
2-benzamido-4,5-D_2-norborneol	^2H		50
D_2			
Ortho-hydrogen	^2H		200
Para- hydrogen			3910
AsH_3	^{75}As	(0.29)	50–100×10^{-3}

Since in mobile liquids and solutions the molecular correlation times τ_c are between 10^{-11} and 10^{-12} s, Equation (4.24) shows that the $T_1(Q)$ relaxation times are primarily dictated by NQCC values.

Finally, it must be emphasized that, in spite of the principal difference between quadrupolar and dipolar coupling, the T_1 behavior in both cases is very similar. In fact, $T_1(Q)$ times are independent of the working frequencies of NMR spectrometers in high-temperature regions. Minimal $T_1(Q)$ times increase proportionally to the Larmor frequency. Note, however, that in contrast to nuclei with spins $I = 1/2$, relaxing by different mechanisms, the relaxation of quadrupolar nuclei is always dominated by quadrupolar interactions, particularly in asymmetric nuclear environments.

Bibliography

1. A. Abragam, *Principles of Nuclear Magnetism*. Oxford at the Clarendon Press: New York, 1985.

2. D. G. Gusev, R. L. Kuhlman, K. B. Renkema, O. Eisenstein, K. G. Caulton. *Inorganic Chemistry* 1996; **35**: 6775.

3. V. I. Bakhmutov, E. V. Vorontsov. *Reviews of Inorganic Chemistry* 1998; **18**: 183.

4. N. Bloembergen, E. M. Purcell, R. V. Pound. *Physical Review*, 1948; **73**: 679.

5. J. Noggle, R. E. Schirmer. *NOE*. Academic Press: London, 1971.

6. L. Sturz, A. Dolle. *Journal of Physical Chemistry A* 2001; **105**: 5055.

7. H. Friebolin. *Basic One- and Two-Dimensional NMR Spectroscopy*. VCH: Weinheim, 1991.

8. V. I. Bakhmutov, E. V. Vorontsov, G. I. Nikonov, D. A. Lemenovskii. *Inorganic Chemistry* 1998; **37**: 279.

9. J. Peralta-Cruz, V. I. Bakhmutov, A. Ariza-Castolo. *Magnetic Resonance in Chemistry* 2001; **39**: 301.

10. B. K. Hunter. *Nuclear Magnetic Resonance Spectroscopy*. Oxford University Press: Oxford, New York, 1993.

11. J. A. S. Smith. *Chemical Society Reviews* 1986; **15**: 225.

12. C. T. Farrar. *An Introduction to Pulse NMR Spectroscopy*. Farragut Press: Chicago, 1987.

13. V. I. Bakhmutov. In: R. Poli, M. Peruzzini (eds). *Deuterium NMR Relaxation as a Method for the Characterization and Study of Transition Metal Hydride Systems in Solution*, in *Recent Advances in Hydride Chemistry*. Elsevier: Amsterdam, London, New York, Paris, Tokyo, 2001; 375–390.

5 Relaxation by Chemical Shift Anisotropy, Spin–Rotation Relaxation, Scalar Relaxation of the Second Kind and Cross-mechanisms

This chapter concerns the mechanisms of nuclear relaxation, which are not associated directly with molecular structures. Nevertheless studies of relaxation by chemical shift anisotropy and scalar relaxation of the second kind allow one to obtain important NMR parameters ($\Delta\sigma$- and J-constants) which are usually masked in NMR spectra. In turn, these spectral parameters depend on electronic structures of investigated compounds. In addition, the mechanisms considered in this chapter can give remarkable contributions to relaxation rates studied for structural reasons. It is obvious that knowledge of the factors controlling these mechanisms helps one to evaluate, for example, dipolar relaxation contributions, used for structural conclusions.

5.1 Relaxation by Chemical Shift Anisotropy

The chemical shift δ in NMR spectra is a measure of *electron screening* of nuclei, placed in an external magnetic field B_0. For this reason, the chemical

Practical NMR Relaxation for Chemists   Vladimir I. Bakhmutov
© 2004 John Wiley & Sons, Ltd   ISBNs: 0-470-09445-1 (HB); 0-470-09446-X (PB)

shift is often reformulated as a *screening constant* σ:

$$\sigma = -\delta \tag{5.1}$$

Electrons, surrounding nuclei, circulate by action of the applied external magnetic field B_0, creating a new magnetic field B_E. The direction of this field is opposite to the field B_0. As a result, a local magnetic field B_{Loc}, is produced:

$$B_{Loc} = B_0 - B_E \tag{5.2}$$

It is obvious that, under these conditions, the NMR signal can be registered if the irradiating frequency is shifted.

According to the theory of electronic screening, the constant σ is a sum of three principal terms: the diamagnetic term σ^{local}(dia), connected with the circulation of unperturbed spherical (s) electrons, the paramagnetic term σ^{local}(para), resulting from perturbed nonspherical electrons and the term σ^*, caused by the magnetic anisotropy of neighboring groups [1]. Combinations of these terms lead to large variations of chemical shifts for different nuclei (Table 5.1).

The σ constants are three-dimensional magnitudes, characterized by σ_{XX}, σ_{YY} and σ_{ZZ} components, shown in Figure 5.1. Usually, these components are not identical, and thus the screening constants are anisotropic. The *chemical shift anisotropy* $\Delta\sigma$ is defined via the expression:

$$\Delta\sigma = \{2\sigma_{ZZ} - (\sigma_{XX} + \sigma_{YY})\}/3 \tag{5.3}$$

In mobile liquids or solutions, the high-amplitude molecular motions average the σ_{XX}, σ_{YY} and σ_{ZZ} components, providing measurements of only *isotropic* screening constants σ (ISO) (or *isotropic* chemical shifts δ (ISO)):

$$\sigma(ISO) = (1/3)(\sigma_{XX} + \sigma_{YY} + \sigma_{ZZ}) \tag{5.4}$$

Table 5.1 Ranges of chemical shifts (SCS) for different nuclei [2, 3]

Nucleus	SCS (ppm)*
^{1}H	$-50-18^{**}$
^{11}B	$-130-95$
^{13}C	$-300-250$
^{19}F	$-250-400$
^{31}P	$-125-500$
^{15}N	$-500-850$
^{195}Pt	$-1370-12000$

*Positive values correspond to a high field displacement
**Including transition and nontransition metal hydrides

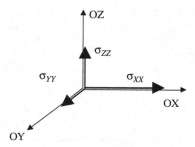

Figure 5.1 The σ_{ZZ}, σ_{XX} and σ_{YY} components of three-dimensional screening constants σ

In solids, molecular motions are slow or very restricted. Therefore, the σ_{XX}, σ_{YY}, σ_{ZZ} components and also chemical shift anisotropies can be calculated directly from static NMR patterns [4] dependent on the symmetry of screening tensors. Figure 5.2 illustrates a solid-state NMR signal, typical of the axially symmetrical screening tensor where:

$$\sigma_{XX} = \sigma_{YY} \neq \sigma_{ZZ} \tag{5.5}$$

When the tensor of magnetic screening has a lower symmetry (i.e. $\sigma_{XX} \neq \sigma_{YY} \neq \sigma_{ZZ}$) lineshapes become more complex.

Chemical shift tensors and $\Delta\sigma$ values, determined for ^{14}N and ^{19}F nuclei in the solid state, are shown in Table 5.2 [3]. It is seen that chemical shift anisotropies of these nuclei are significant, and even comparable with the scales of chemical shifts (see Table 5.1). This is not a rule, but it provides

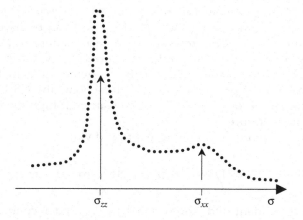

Figure 5.2 Lineshape of a solid-state NMR resonance with an axially symmetrical screening tensor: $\sigma_{ZZ} \neq \sigma_{XX} = \sigma_{YY}$. The resonances are typically observed for C=O groups in the static solid state $^{13}C(^{1}H)$ NMR spectra

Table 5.2 Components of screening tensors and the $\Delta\sigma$ values (ppm) for some nuclei, measured in the solid state

Compound	σ_{ZZ}	σ_{YY}	σ_{XX}	$\Delta\sigma$
$^{14}NH_3$	237.3	278.0	278.0	−44.3
$Me^{14}NC$	370	10	10	360
$HC^{14}N$	348	−215	−215	563
$C_6{}^{19}F_6$	465.9	310.8	310.8	155.1
$Xe^{19}F_4$	528.5	58.5	58.5	470

an estimate of the role of chemical shift anisotropy in relaxation of heavy nuclei. For example, the $\Delta\sigma$ values of carbonyl groups reach 200 ppm and their influence on ^{13}C relaxation is obvious.

Random molecular motions cause fluctuations of local magnetic fields, created by chemical shift anisotropies, and then the CSA relaxation rates, $1/T_1(CSA)$, is:

$$1/T_1(CSA) = (1/15)\gamma_I^2 B_0^2 (\Delta\sigma)^2 [2\tau_c/(1 + \omega_I^2 \tau_c^2)] \qquad (5.6)$$

Simple estimations show that this mechanism is particularly effective for such nuclei as ^{13}C, ^{15}N, ^{19}F and ^{31}P, having the large CSA values. However, at the extremely high magnetic fields, used in modern NMR experiments, this mechanism can become important, even for protons with relatively low $\Delta\sigma$ values (~ 20 ppm).

According to Equation (5.6), under conditions of fast molecular motion (i. e. $\omega^2 \tau_c^2 \ll 1$) the CSA nuclear relaxation is *field dependent* and $T_1(CSA)$ times decrease with increasing the B_0^2. This effect can be used as a good test for the presence of CSA relaxation. Table 5.3 demonstrates the field-dependent spin–lattice relaxation rates measured for ^{13}C nuclei in fullerene molecules C_{60} in chlorobenzene-d_5 [3]. It is obvious that in the absence of protons, as sources of dipole–dipole interactions, the CSA mechanism becomes dominant, providing >67% of relaxation at high magnetic fields, even at high temperatures.

The CSA spin–spin relaxation rate, expressed via:

$$1/T_2(CSA) = (1/90)\gamma_I^2 B_0^2 (\Delta\sigma)^2 [8\tau_c + 6\tau_c/(1 + \omega_I^2 \tau_c^2)] \qquad (5.7)$$

is also field dependent and proportional to B_0^2 under conditions of fast molecular motion. Combining Equations (5.6) and (5.7) leads to a ratio: $T_1(CSA)/T_2(CSA) = 7/6$. This ratio, obtained experimentally, is an additional test for domination of the CSA mechanism in nuclear relaxation.

Table 5.3 Field-dependent spin–lattice relaxation rates of ^{13}C nuclei in a chlo-robenzene-d$_5$ solution of fullerene C$_{60}$. R_1, R_1^{CSA} and R_1^{SR} are total, CSA and SR relaxation rates, respectively (s^{-1})

T(K)	$R_1 \times 10^2$ 9.4/4.7 T	$R_1^{CSA} \times 10^2$ 9.4/4.7 T	$R_1^{SR} \times 10^3$ 9.4/4.7 T	CSA(%) 9.4/4.7 T	SR(%) 9.4/4.7 T
273	2.44/6.22	2.42/6.06	0.16/0.16	99/97	1/3
288	2.00/6.19	1.84/4.60	1.59/1.59	92/74	8/26
303	1.72/5.79	1.52/3.80	1.99/1.99	88/66	12/34
318	1.47/5.36	1.25/3.11	2.25/2.25	85/58	15/42
333	1.23/6.16	0.82/2.05	4.11/4.11	67/33	33/67

5.2 Spin–Rotation Relaxation

The spin-rotation mechanism (SR) makes a remarkable contribution to nuclear relaxation in relatively small molecules, undergoing fast rotations in *nonviscous* media. Note that electrons in the molecules create magnetic moments, even in the absence of external magnetic fields. Then, fast rotational molecular reorientations result in the spin–rotation relaxation rate:

$$1/T_1(SR) = I_r^2 C^2 / 9\hbar^2 \, \tau_c \qquad (5.8)$$

where I_r is the molecular inertia moment, C is the spin–rotation constant, measured in frequency units, and τ_c is the molecular motion correlation time. It must be emphasize that this simple relationship is valid only for spherical molecules. Equation (5.8) shows that the effectiveness of the SR mechanism depends on C values. These values are significant for heavy nuclei (^{31}P, ^{19}F etc). For example, in the case of ^{19}F nuclei, the C constants can reach 2000 Hz.

Inspection of Equation (5.8) reveals a unique feature of the spin–rotation mechanism: T_1(*SR*) *times decrease with increasing the temperature*. This effect is a good test for the presence of SR relaxation.

As we have shown in Chapter 4, proton–phosphorus dipole–dipole inter-actions do not affect the ^{31}P T_1 times measured for the molecules shown in Figure 4.7. Experiments at magnetic fields of 6.3 and 9.3 T do not reveal the presence of a contribution of the CSA mechanism because the ^{31}P T_1 times are independent of the external magnetic field. At the same time, on heating, the ^{31}P T_1 values reduce (Table 5.4) and hence the spin–rotation mechanism dominates phosphorous relaxation [5].

As it has been mentioned, in the absence of intensive dipole–dipole interactions in the fullerene molecules C$_{60}$, the CSA mechanism dominates ^{13}C T_1 relaxation at a magnetic field of 9.4 T (Table 5.3). However, the situation changes at lower magnetic field (4.7 T) and higher temperatures. According to the theory, the SR contribution increases with the temperature and, for example, at 333 K this mechanism provides 67% of the ^{13}C T_1 relaxation rate.

Table 5.4 Total ^{31}P T_1 times, measured in benzene-d_6 solutions of compounds 1–5 (see Figure 4.7)

Compound	^{31}P T_1 (s) 295/338 K
1	$10.2 \pm 0.1 / 6.8 \pm 0.3$
2	$11.8 \pm 0.1 / 10.5 \pm 0.1$
3	$11.5 \pm 0.3 / 8.6 \pm 0.3$
4	$10.7 \pm 0.1 / 8.3 \pm 0.2$
5	$11.1 \pm 0.1 / 7.7 \pm 0.2$

5.3 Interference Mechanisms of Nuclear Relaxation

By definition, relaxation rates are additive magnitudes. When various relaxation mechanisms operate simultaneously (for example, the dipole–dipole, chemical shift anisotropy and spin–rotation pathways), the corresponding contributions should be summarized to give the total relaxation rates, $1/T_1$(tot):

$$1/T_1(\text{tot}) = 1/T_1(\text{DDI}) + 1/T_1(\text{CSA}) + 1/T_1(\text{SR}) \quad (5.9)$$

All the contributions in Equation (5.9) are independent and can be separated by the appropriate spectral or chemical procedures (see below). However, two mechanisms of different nature can undergo interference in the so-called cross-correlation mechanism. Such interference is often observed for the dipole–dipole and CSA interactions. For example, the cross-correlation between the ^{15}N CSA and ^{15}N–^{1}H dipole–dipole interactions leads to differential transverse relaxation, measured for two components of ^{15}N doublets [6].

Theoretically, the interference relaxation pathway can operate, even in a pair of protons. If these protons are separated by internuclear distance r(H–H) and molecular tumbling is fast, the cross-relaxation term, $1/T_1$(DDI, CSA), is expressed as a function of the dipolar relaxation rate, R(DDI) (equal to $1/T_1$(DDI):

$$1/T_1(\text{DDI, CSA}) = -(4\pi/5\sqrt{3})(3\cos^2 \Theta - 1)(\nu_0 \Delta\sigma)$$
$$\times R(\text{DDI})/\{(\mu_0/4\pi)\gamma_H^2\hbar/r(\text{H–H})^3\} \quad (5.10)$$

where ν_0, μ_0, γ_H, $\hbar$ and $\Delta\sigma$ are the known physical parameters and Θ is the angle between the r(H–H) vector and the principal axis of the chemical shift tensor [7] (see Figure 5.3). It is easy to show that when the principal axis is perpendicular to the r(H–H) vector and r(H–H) = 2.4 Å, $\Delta\sigma$ = 26 ppm and

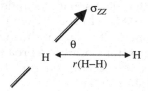

Figure 5.3 Angle Θ between the internuclear H–H vector and the principal axis of the chemical shift tensor, σ_{ZZ}, determining values of relaxation contributions from interference of the dipole–dipole and CSA mechanisms

$\nu_0 = 400$ MHz, Equation (5.10) converts to the ratio:

$$1/T_1(\text{DDI, CSA}) = 0.3\, R(\text{DDI}) \tag{5.11}$$

Thus, the DDI/CSA interference contribution can reach 30% of the dipole–dipole relaxation rate at strong magnetic fields and big $\Delta\sigma$ values. Note that the presence of the interference DDI/CSA terms cannot be established experimentally by standard inversion recovery experiments. In such cases, one can recommend application of pulse sequences [8]:

$$180°-\tau-20° \tag{5.12}$$

5.4 Scalar Relaxation of the Second Kind

Scalar interactions affect the spin–spin relaxation of A nuclei, which are coupled by X nuclei, when the A–X coupling is time dependent. Phenomenologically, scalar relaxation of the second kind (SRSK) appears as an additional broadening of NMR signals. Figure 5.4 demonstrates a typical ^{1}H resonance, observed usually for HNR$_2$ or H$_2$NR amino groups. In accordance

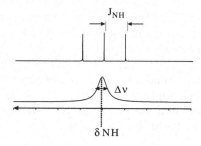

Figure 5.4 The shape of a ^{1}H NMR signal, observed for HNR$_2$ or H$_2$NR groups in the absence (upper part) and presence (lower part) of proton scalar relaxation of the second kind

with the well-known rule, the ^{1}H resonance should show a 1:1:1 triplet with $^1J(^1H-^{14}N)$ constant of $\simeq$ 60 Hz. However, owing to fast relaxation of ^{14}N nuclei (as fast flip–flop motions of ^{14}N spins during a detection period at the proton frequency) this triplet transforms to a broadened singlet.

Usually, the linewidth $\Delta\nu^{SRSK}$ for a resonance of A nuclei, coupled with X nuclei, depends on the T_2(SRSK) time:

$$\Delta\nu^{SRSK} = 1/\pi \, T_2(\text{SRSK}) \tag{5.13}$$

where the rate of spin–spin relaxation is:

$$1/T_2(\text{SRSK}) = (8/3)\pi^2 J^2(\text{AX}) \times I_X(I_X + 1)T_{1X}/\{1 + (\omega_X - \omega_A)^2 T_{1X}^2\} \tag{5.14}$$

Thus, the spin–spin coupling constants $J(\text{AX})$ are 'masked' by linewidths, controlled, in turn, by the spin–lattice relaxation times $T_1(X)$. It is obvious that these masked $J(\text{AX})$ constants can be determined if the T_{1X} values are known, and vice versa.

The SRSK mechanism plays an important role in relaxation of nuclei, coupled with quadrupolar nuclei, for example, in $^1H-^{11}B$, $^{13}C-^{14}N$ and $^{31}P-^{14}N$ pairs, where ^{1}H, ^{13}C and ^{31}P are target nuclei. The last case is illustrated by the data in Table 4.1. As it is seen, the ^{31}P T_2 times are strongly shorter than ^{31}P T_1. Since in solutions $T_1 = T_2$, a $T_2 < T_1$ ratio directly shows the presence of the SRSK mechanism. The contribution of this mechanism can be obtained by:

$$1/T_2(\text{SRSK}) = 1/T_1 - 1/T_2 \tag{5.15}$$

According to the data in Table 4.1, scalar interactions completely dominate the ^{31}P spin–spin relaxation.

The hydride ligands in complexes 1–3 [9] (Figure 5.5) are adjacent to the rapidly relaxing ^{93}Nb quadrupolar nuclei, which could cause scalar coupling. In accord, the hydride resonances are strongly broadened (Table 5.5) and the observed linewidths, $\Delta\nu^{obs}$, do not correspond to T_1 times. In fact, $\Delta\nu^{obs} > 1/\pi T_1$. Moreover, the $\Delta\nu^{obs}$ values increase with the temperature. Note that this effect directly follows from Equation (5.14) at $1 \gg (\omega_{Nb} - \omega_H)^2 T_{1Nb}^2$. The SRSK $\Delta\nu^{Nb-H}$ broadenings, caused by the ^{93}Nb$-^1$H coupling, can be calculated as $\Delta\nu^{Nb-H} = \Delta\nu^{obs} - \Delta\nu$, where $\Delta\nu = 1/\pi T_1$. It is obvious that the H(A) resonances are broadened to a greater extent, owing to the larger $^1J(^1H-^{93}Nb)$ spin–spin coupling constants. Then, the $^1J(^1H-^{93}Nb)$ values (Figure 5.5) can be estimated via Equation (5.14) where the T_{1X} time (equal to $T_1(^{93}Nb)$) takes a value of 2×10^{-5} s measured for Cp_2NbH_3 at room temperature.

Finally, it should be emphasized that the broadening effects caused by the SRSK mechanism can be successfully used for line assignments in NMR spectra.

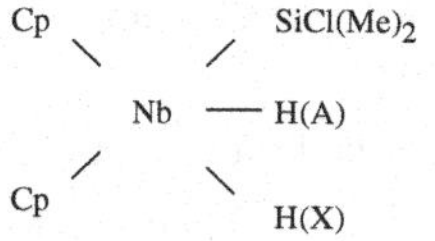

1

$\delta(H^A) = -4.47$ ppm
$\delta(H^X) = -4.27$ ppm
$J(H^X\text{-}Nb) = 46$ Hz
$J(H^A\text{-}Nb) = 109$ Hz

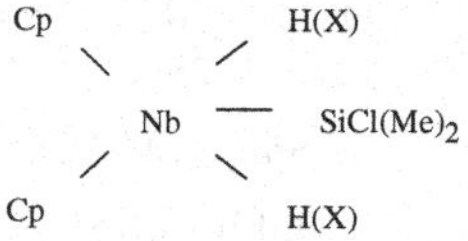

2

$\delta(H^X) = -4.64$ ppm
$J(H^X\text{-}Nb) = 72$ Hz

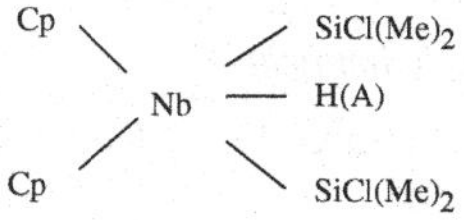

3

$\delta(H^A) = -5.11$ ppm
$J(H^A\text{-}Nb) = 126$ Hz

Figure 5.5 Structures and ^{1}H NMR parameters of Nb hydride complexes: the $^1J(^1H-^{93}Nb)$ spin–spin coupling constants are determined on the basis of linewidths and T_1 times measured for the hydride resonances (Table 5.5)

Table 5.5 Temperature dependences of line widths $\Delta\nu$ (Hz) and relaxation times T_1 (s), measured in toluene-d_8 at 400 MHz, for the hydride resonances in Nb complexes 1–3 (Figure 5.5)

Compound	$\Delta\nu^{obs}(H^X)/\Delta\nu^{Nb-H}(H^X)$	$\Delta\nu^{obs}(H^A)/\Delta\nu^{Nb-H}(H^A)$	T_1(HX)/T_1(HA)	T(K)
1	12.4	Very broadened		340
	4.8/4.4	25.0/24.7	0.903/0.954	294
	4.4/3.1	11.0/9.8	0.241/0.269	240
	5.2/2.4	7.0/4.5	0.113/0.126	220
2	26.0			340
	11.2/10.9		1.368	294
	5.2/4.2		0.368	240
	4.7/2.9		0.178	220
	4.0/1.4		0.125	200
3		32.6/32.3	1.274	298
		23.6/23.0	0.174	220

Bibliography

1. B. K. Hunter. *Nuclear Magnetic Resonance spectroscopy*. Oxford University Press: Oxford, New York, 1993.
2. J. B. Lambert, F. G. Riddel. *The Multinuclear Approach to NMR Spectroscopy*, by Ed. Reidel D, Publishing Company, Boston, 1982.

3. N. H. Martin, M. H. Issa, R. A. McIntyre, A. A. Rodriguez. *Journal of Physical Chemistry A* 2000; **104**: 11278.

4. M. J. Duer (ed.) *Solid State NMR Spectroscopy* Blackwell: Oxford, 2002.

5. J. Peralta-Cruz, V. I. Bakhmutov, A. Ariza-Castolo. *Magnetic Resonance in Chemistry* 2001; **39**: 301.

6. C. Wang, A. G. Palmer. *Magnetic Resonance Chemistry* 2003; **41**: 866.

7. S. Aime, W. Dastru, R. Gobetto, A. Viale. NMR relaxation studies of polynuclear hydride derivatives. In: R. Poli, M. Peruzzini (eds), *Recent Advances in Hydride Chemistry* 2001; 351–374.

8. C. Dalvit, C. Bodenhausen. *Chemical Physics Letters* 1989; **161**: 554.

9. V. I. Bakhmutov, J. A. K. Howard, D. A. Kenn *et al.* *Journal of the Chemical Society Dalton Transaction* 2000; 1631.

6 Nuclear Relaxation in Molecular Systems with Anisotropic Motions

Nuclear relaxation is a direct way of understanding dynamic processes in condensed matter at a molecular level, and of deducing the basic connectivity between primary structure and molecular motions, and hence the mechanical properties of molecular systems. For these reasons, this method is widely used in spite of nontrivial separations of relaxation contributions from translational, rotational and internal motions in liquids or even in the solid state. In addition, nuclear relaxation is a non-perturbing method for studying molecular reorientations because the target nuclei usually occur naturally in the compounds investigated (for example, ^1H, ^{19}F, ^{31}P) or they can be substituted for ^2H, ^{13}C or ^{15}N nuclei without perturbation of the dynamics.

The frequency scale of motions available to investigators, as a function of the magnetic field strength, is very large, from 10^2 Hz (slow motions) to 10^{12} Hz (fast motions) [1]. It is obvious that rates of motions depend on the nature of molecules, their size and the physical properties of the medium. For example, polymer systems show the widest spectrum of molecular motions. Local reorientations in such systems occur at frequencies of the order of 10^9 Hz via conformational transitions. Internal rotations of attached groups occur at frequencies 10^9-10^{10} Hz. Restricted librational motions with amplitudes of $30°-60°$ occur at frequencies of $10^{11}-10^{12}$ Hz and, finally, overall molecular tumbling occurs at frequencies of 10^5-10^7 Hz.

Practical NMR Relaxation for Chemists   Vladimir I. Bakhmutov
© 2004 John Wiley & Sons, Ltd   ISBNs: 0-470-09445-1 (HB); 0-470-09446-X (PB)

Molecular motions are not the principal topic of this book. Nevertheless, we use different theoretical approaches, developed for an adequate description of molecular reorientations, in order to show how character of motions affects relaxation times measured experimentally. Mathematically the influence of the molecular motions on times of nuclear relaxation is expressed via the spectral density function $J(\omega, \tau_C)$. If molecular motions are fully isotropic, as in the case of spherical molecules, the $J(\omega, \tau_C)$ function takes the Bloembergen–Purcell–Pounds' form (see Equations (4.4), (4.5) and (4.23), corresponding to the symmetric V-shaped plots of $\ln(T_1)$ versus $1/T$. However, motions of real molecules are anisotropic and complex, particularly in the case of polymer systems. In this situation, description of motions requires modifications of the spectral density functions $J(\omega, \tau_C)$ which are illustrated in this chapter with the help of the simplest anisotropic motional models.

6.1 Spin–Lattice Nuclear Relaxation in Ellipsoidal Molecules: Temperature Dependences of T_1 Times

Rotational molecular reorientations in solutions of many nonspherical molecules are similar to those of a symmetrical ellipsoid in a continuous medium (Figure 6.1). Ellipsoidal molecules have two principal rotational directions with different moments of inertia. For this reason, their motions are characterized by two molecular motion correlation times $\tau_\perp$ and $\tau_\parallel$ (or the corresponding diffusion coefficients) and two activation energies. A good chemical example of such molecules is the bipyramidal hydride complex, $IrH_5(PPr^i{}_3)_2$, with two *trans*-located bulky phosphorous ligands. The simple calculation shows that this molecule has two inertia moments in the ratio 0.41:1 [2].

Woessner has carried out a detailed analysis of the ^1H dipole–dipole spin–lattice relaxation in an idealized case, when two protons are fixed in a symmetric ellipsoid undergoing a thermally activated anisotropic rotational diffusion [3, 4]. It has been shown that the spectral density function of Bloembergen, Purcell and Pound (see Equation 4.5) still describes adequately the relaxation behavior of this system if the isotropic motional correlation times τ_C are replaced by the, so-called, effective correlation times τ_{eff}, given by the expression:

$$\tau_{eff} = A(\theta)\,\tau_A + B(\theta)\,\tau_B + C(\theta)\,\tau_C \tag{6.1}$$

Coefficients A, B and C depend on the angle, θ, formed by the vector connecting the protons and the rotational axis (Figure 6.1):

$$A(\theta) = (1/4)(3\cos^2\theta - 1)^2$$

$$B(\theta) = 3\sin^2\theta\cos^2\theta \tag{6.2}$$

$$C(\theta) = (3/4)\sin^4\theta$$

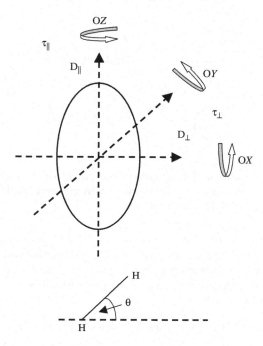

Figure 6.1 Rotational reorientation of an ellipsoidal molecule in a continuous medium with the correlation times $\tau_\perp$ and $\tau_\parallel$. When two protons are fixed in such an ellipsoid, the orientation of the dipolar H—H vector is characterized by the angle θ formed by the vector and one of the rotational axis (dashed line)

In turn, the correlation times τ_A, τ_B and τ_C in Equation (6.1) are given by:

$$\tau_A = \tau_\perp$$
$$(\tau_B)^{-1} = (5/6)\tau_\perp^{-1} + (1/6)\tau_\parallel^{-1} \qquad (6.3)$$
$$(\tau_C)^{-1} = (1/3)\tau_\perp^{-1} + (3/2)\tau_\parallel^{-1}$$

Thus, when the structure of an ellipsoidal molecule is known, measurements of T_1 times for different target nuclei, located in this molecule, finally give the $\tau_\perp$ and $\tau_\parallel$ values.

Now we consider a pair of protons, fixed in an ellipsoidal molecule, and demonstrate how the motional anisotropy affects their spin–lattice relaxation. The anisotropy of motions of ellipsoidal molecules can be quantitatively characterized by parameter ρ, expressed by:

$$\rho = D_\parallel / D_\perp \qquad (6.4)$$

where $D_\parallel$ and $D_\perp$ are the rotational diffusion constants, corresponding to the $\parallel$ and $\perp$ axes, respectively. Take, for simplicity, the angle θ equal to

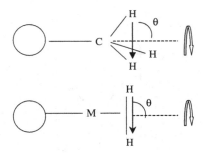

Figure 6.2 Fast internal rotation (rotational diffusion) of CH_3 groups in organic molecules and (H_2) ligands in transition metal dihydrogen complexes, leading to a four-fold elongation of ^1H T_1 relaxation times

90°. This situation occurs, for example, in methyl groups or dihydrogen ligands in Figure 6.2, the protons of which participate simultaneously in molecular tumbling and intramolecular rotation. According to Bakhmutov and Gusev [5], under these conditions, the rate of dipolar relaxation, $1/T_1$, can be expressed by:

$$1/T_1(H{-}H) = (3/40)(\mu_0/4\pi)^2\gamma_H{}^4\hbar^2 r(H{-}H)^{-6}\tau_c$$
$$\{1/(1 + \omega_H{}^2\tau_c{}^2) + 4/(1 + 4\omega_H{}^2\tau_c{}^2) + 3a/(a^2 + \omega_H{}^2\tau_c{}^2)$$
$$+ 12a/(a^2 + 4\omega_H{}^2\tau_c{}^2)\} \tag{6.5}$$

where $a = (2\rho + 1)/3$ and $\tau_C = 1/6\,D_\perp$. It is easy to show that at $\rho = 1$, Equation (6.5) converts to the Bloembergen–Purcell–Pounds function, i.e. to the case of isotropic molecular motions. On the other hand when $\rho \to \infty$ and internal rotation becomes very fast on the time scale of molecular tumbling, the function $J(\omega_H, \tau_c)$ in Equation (6.5) transforms to the Woessners' form:

$$J(\omega_H, \tau_c) = \{(3\cos^2\theta - 1)^2/4\}(\tau_c/(1 + \omega_H{}^2\tau_c{}^2) + 4\tau_c/(1 + 4\omega_H{}^2\tau_c{}^2) \tag{6.6}$$

where at $\theta = 90°$ the factor $(3\cos^2\theta - 1)^2$ is equal to 1. Take $r(H{-}H) = 2$ Å and calculate via Equation (6.5) the ^1H T_1 time as a function of τ_C. The results, obtained for ρ varying from 1 to 2, 5, 10, 50 and ∞ and at the NMR frequency of 200 MHz, are shown in Figure 6.3 as plots of the $\ln(T_1)$ versus τ_C. Since the correlation time τ_C depends on the temperature (see Equation 1.13), the curves in Figure 6.3 reproduce the temperature dependences of relaxation times at increasing anisotropy of the motion. Inspection of these data allows one to formulate the following important conclusions.

First, increasing the anisotropy ρ causes an increase of ^1H T_1 times in spite of the constant proton–proton distance; the increase is particularly strong for the left and right wings of the curves corresponding to high- and low-temperature regions, respectively.

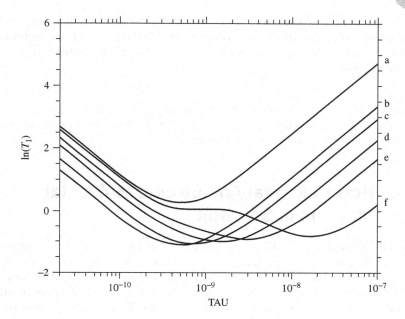

Figure 6.3 Plots of $\ln(T_1)$ versus τ_C (equivalent to T_1 temperature dependences) calculated by Equation (6.5) for two protons, separated by 2 Å and resonating at 200 MHz. The calculations were carried out over a range of ρ values from ∞ to 50, 10, 5, 2 and 1 (in the high-temperature region from top to bottom). (Reproduced from D. G. Gusev, D. Nietlispach, A. B. Vymenits, V. I. Bakhmutov, H. Berke. *Inorganic Chemistry*, 1993; **32**: 3270, with permission from Elsevier)

Second, anisotropic reorientations with intermediate ρ parameters (see $\rho = 10$ and 50) result in the appearance of the nonsymmetric plots with respect to τ_C values, corresponding to T_1 $_{min}$ times. Moreover, at $\rho = 50$ the $\ln(T_1)$ plot has two T_1 minima, showing directly the presence of two different motions. However if the anisotropy of molecular reorientations is low (see $\rho = 2$–5) or very high ($\rho = \infty$), these features are not observed. Hence, such reorientations and fully isotropic motions are indistinguishable phenomenologically.

Third, when the ρ value increases from 1 to 50, T_1 $_{min}$ times are shifted towards lower temperatures.

Fourth, all the calculated $T_1(\tau_C)$ curves show identical slopes of their left and right wings, corresponding to high- and low-temperature regions, respectively.

Thus treatment of the variable-temperature T_1 relaxation curves, collected experimentally for ellipsoidal molecular systems, under the assumption of isotropic motions, leads to *effectively elongated* internuclear distances. It is obvious that errors in the distance calculations depend on the character of the anisotropy of the motion. They become maximal when one of the motions of ellipsoid molecules is very fast and the dipolar H−H vectors are perpendicular to the internal motional axis. Actually the latter leads to the situations when:

$$T_1(\text{aniso}) = 4\,T_1(\text{iso}) \tag{6.7}$$

However, it is remarkable that activation energies E_a, obtained in the anisotropic approximation, remain meaningful while the correlation time constants τ_0 become fictitious parameters.

6.2 How to Reveal Anisotropic Molecular Motions in Solution

As has been mentioned in Chapter 2, strongly anisotropic molecular motions can lead to a nonexponential nuclear relaxation. It is obvious that in such cases, the data collected, for example, by inversion recovery experiments, are poorly approximated by the standard *monoexponential* fitting procedures. The latter, in turn, directly reveal the presence of anisotropic motions. In many cases, T_1 times can be obtained from the nonexponential NMR decays as the initial slopes R of the collected inversion recovery curves. According to Daragan *et al.* [6], these slopes are determined by calculations of the inversion recovery curves in the parabolic approximation $\ln I(t) = I_0 - Rt - At^2$. These treatments can be carried out with the help of the program RECULVE in NMR software packages. In practice, however, relaxation curves are often monoexponential even in the presence of anisotropic motions. In some cases (as we have shown), anisotropic motions are easily revealed when plots of $\ln(T_1)$ versus $1/T$ deviate from the symmetric V-shaped curves or show additional T_1 minima. In the absence of such features, the anisotropy of molecular motions can be established by relaxation measurements carried out for different nuclei in the same molecule. Figure 6.4 summarizes the results of relaxation experiments on the ^{1}H resonances of the octahedral

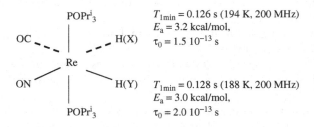

Figure 6.4 Results of ^{1}H T_1 variable-temperature experiments carried out for the H(X) and H(Y) ligands in the dihydride Re complex in toluene-d_8

Re dihydride. It must be emphasized that the H(X) and H(Y) resonances show monoexponential ^{1}H T_1 behavior over a large temperature range. Moreover, their $T_{1\,min}$ times are very similar, as well as the E_a values calculated from the T_1 curves. However, in the absence of visible geometrical reasons, the ^{1}H $T_{1\,min}$ times of the H(X) and H(Y) igands are observed at different temperatures: 194 versus 188 K for the H(X) and H(Y) resonances, respectively [5]. A similar situation is shown in Figure 6.5 where the location of the $T_{1\,min}(CH_3)$ for complex $ReH(CO)_3(POPr^i_3)_2$ is markedly shifted to a lower temperature with respect to a minimum observed for the hydride ligands. Thus, in spite of a monoexponential relaxation, the motions of $ReH(CO)_3(POPr^i_3)_2$ and $ReH_2(CO)(NO)(POPr^i_3)_2$ are anisotropic.

High-temperature T_1 experiments on nuclei of different nature can also reveal anisotropic motions. In high-temperature regions, the plots of $\ln(T_1)$ versus $1/T$ are linear and their slopes correspond to activation energies of molecular motions. If molecular reorientations are isotropic, then different target nuclei show identical E_a values. In contrast, variable-temperature T_1 experiments on ^{1}H, ^{31}P and ^{187}Re nuclei in a toluene-d_8 solution of complex

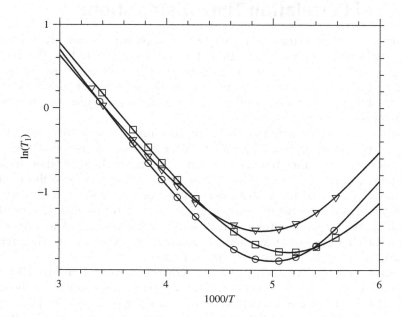

Figure 6.5 Variable-temperature ^{1}H T_1 data collected in a toluene–d_8 solution of complex $Re(CO)_3H(POPr^i_3)_2$: triangles H ligand at 200 MHz; circles H ligand at 300 MHz; squares CH_3 groups at 300 MHz ($T_{1\,min}$ for the CH_3 resonance is not observed at 200 MHz). (Reproduced from D. G. Gusev, D. Nietlispach, A. B. Vymenits, V. I. Bakhmutov, H. Berke. *Inorganic Chemistry* 1993; **32**: 3270, with permission from Elsevier)

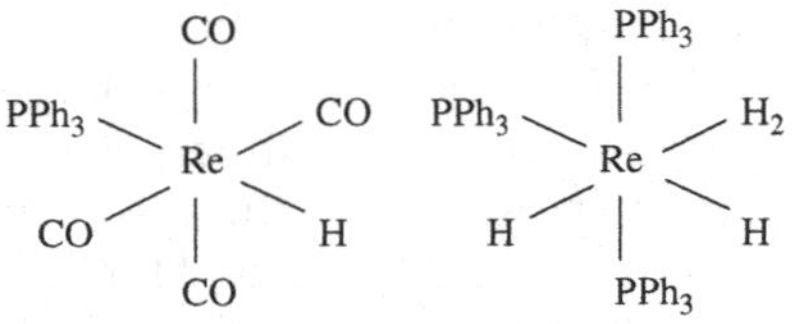

$$
\begin{array}{ccccc}
 & \text{CO} & & & \text{PPh}_3 \\
\text{PPh}_3 & | & \text{CO} & \text{PPh}_3 & | & \text{H}_2 \\
 & \text{Re} & & & \text{Re} \\
\text{CO} & | & \text{H} & \text{H} & | & \text{H} \\
 & \text{CO} & & & \text{PPh}_3 \\
 & \mathbf{1} & & & \mathbf{2}
\end{array}
$$

Figure 6.6 The octahedral rhenium hydride complexes, undergoing anisotropic rotational reorientations in toluene-d_8 [7]

$ReH(CO)_3PPh_3$ (compound 1 in Figure 6.6) lead to *three* E_a values varying from 2.0 to 3.8 kcal/mol [7]. Hydride complex 2 also shows two E_a values (2.8 ± 0.3 and 3.9 ± 0.3 kcal/mol, toluene-d_8) obtained by the T_1 experiments at ^{1}H and ^{2}H frequencies [7]. Thus, molecular motions are again anisotropic in both cases.

6.3 Nuclear Relaxation in the Presence of Correlation Time Distributions

In solids and viscous liquids, for some reason (which are not always clear), the variable-temperature relaxation data are poorly fitted to the Bloembergen–Purcell–Pounds' theory. In other words, molecular motion is not described by a single correlation time τ_C, or the temperature dependence of the τ_C is not exponential. One of the basic reasons for this phenomenon is the appearance of *correlated* or *cooperative* motions [1].

Usually, one can assume that within an ensemble of reorienting units, there are sub-ensembles, each of which is characterized by a single correlation time τ_C. However, this time changes from one sub-ensemble to another. In such a situation, one talks about *a correlation time distribution* where the τ_C time becomes a center of the distribution with a certain width, depending on the type of motions and the properties of molecular systems. It should be noted that the correlation time distributions can be considered as convenient mathematical expressions applied for treatment of experimental data which deviate from the Bloembergen–Purcell–Pounds' theory.

There are several symmetric and nonsymmetric correlation time distributions, successfully applied for interpretation of nuclear relaxation in the solid state: Cole–Cole, Cole–Davidson, FAN, Fuoss–Kirkwood etc. [1]. Determination of the type of distribution is a very important and nontrivial problem related to details of investigated systems and dynamic processes. In other words, practical applications of the correlation time distributions are addressed to studies of molecular motions when the elementary molecular structure is already known. In the context of structural studies it is important to show how the presence of correlation time distributions perturbs the measured T_1 times. In fact, theoretically, the existence of the reorienting ensembles and sub-ensembles can be proposed, even in regular liquids or solutions becoming viscous at low temperatures.

The symmetric distribution of Fuoss and Kirkwood, is often applied for studies of nuclear relaxation in the solid state. In the presence of the Fuoss–Kirkwood distribution, the spin–lattice relaxation rate, $1/T_1$, in a pair of protons, is:

$$1/T_1(\text{H–H}) = \{DC_{\text{H–H}}\tau_c\beta/\omega_H\tau_c\} \times$$

$$*\{(\tau_c\omega_H)^\beta/(1 + (\omega_H\tau_c)^{2\beta}) + 2\,(2\omega_H\tau_c)^\beta/(1 + (2\omega_H\tau_c)^{2\beta}\} \qquad (6.8)$$

where $DC_{\text{H–H}}$ is the proton–proton dipolar coupling and β is the width of the distribution ($0 < \beta \leqslant 1$). It is seen that at $\beta = 1$ Equation (6.8) converts to the Bloembergen–Purcell–Pounds' expression, i.e. to the isotropic motional model. Figure 6.7 shows the ^{1}H $T_1(\tau_C)$ dependence calculated via Equation (6.8) for a pair of protons, separated by 2 Å, at the NMR frequency of 200 MHz and $\beta = 1$ (isotropic motions), 0.7 and 0.5. In spite of the constant proton–proton distance, decreasing the β value results in an increase of the relaxation times in the ^{1}H $T_1\text{ min}$ region. The slopes of the linear sections, corresponding to fast and slow molecular motions, depend on β. Thus, effective activation energies of molecular motions $E_a(\text{eff})$ are:

$$E_a(\text{eff}) = \beta\,E_a \qquad (6.9)$$

where E_a is a natural activation energy. Comparison of the plots in Figure 6.7 shows that treatment of the Fuoss–Kirkwood T_1 times in the framework of an isotropic motion model (i.e. $\beta = 1$) will lead to effective internuclear

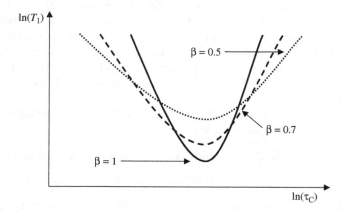

Figure 6.7 Logarithmic plots of T_1 versus τ_C (schematic) obtained for a pair of protons separated by 2 Å at 200 MHz in the presence of Fuoss–Kirkwood correlation time distribution with variation in the width of the distribution β

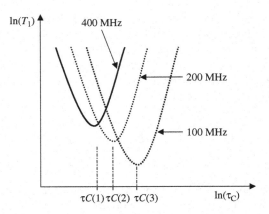

Figure 6.8 Variable-field T_1 data in the presence of Fuoss–Kirkwood correlation time distribution at $\beta = 0.5$ (schematic). The locations of the $T_{1\ min}$ times at different NMR frequencies allow one to determine the τ_{Cmin} values ($\tau\,C(1)$, $\tau\,(C2)$ and $\tau\,C(3)$) corresponding to different temperatures

distances. The distances will be increased in the $T_{1\ min}$ regions and shortened in the high- and low-temperature zones.

By definition, the Bloembergen–Purcell–Pounds and Fuoss–Kirkwood models of nuclear relaxation correspond to the symmetric plots of $\ln(T_1)$ versus $\ln(\tau_C)$ or $1/T$. For this reason, *a priori* these models are indistinguishable. Nevertheless, the Fuoss–Kirkwood distribution can be revealed experimentally by variable-temperature and variable-field T_1 measurements. Figure 6.8 illustrates these experiments schematically as plots of the T_1 versus τ_C in logarithmic coordinates. In accordance with Equation (6.8), the plots go through minima, observed for different magnetic fields at different temperatures. Thus, these data lead to a set of the $\tau_C(T_{1\ min})$ values, which give, in turn, a natural activation energy E_a. An effective activation energy E_{eff} can be obtained as the slopes of the linear sections in the high- and low-temperature regions. It is obvious that at $E_a \neq E_{eff}$ the relaxation data can be treated with the help of the Fuoss–Kirkwood model with the β calculated via Equation (6.9).

In the presence of the Cole–Davidson correlation time distribution [1], the ^{1}H T_1 time for a pair of protons is:

$$1/T_1(H-H) = DC_{H-H}\tau_c \times \{\sin(\beta\arctan \tau_c\omega_H)/\omega_H\tau_c(1 + (\omega_H\tau_c)^2)^{\beta/2})$$

$$+ 4\sin((\beta\arctan 2\tau_c\omega_H)/2\omega_H\tau_c(1 + (2\omega_H\tau_c)^2)^{\beta/2}\} \qquad (6.10)$$

By definition, the Cole–Davidson plots of $\ln(T_1)$ versus $\ln(\tau_c)$ (or $1/T$) are asymmetric relative to $T_{1\ min}$ locations (see Figure 6.9). It is obvious that such a feature, observed experimentally, is a good basis for application of this model. For example, the variable-temperature ^{1}H T_1 times in

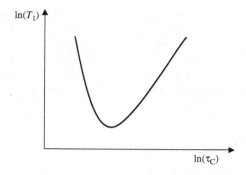

Figure 6.9 Variable-temperature T_1 times in the presence of the asymmetric Cole–Davidson correlation time distribution

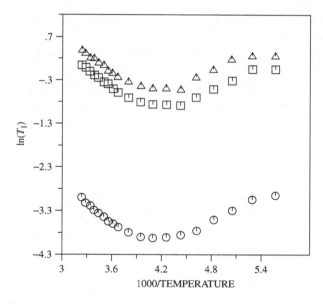

Figure 6.10 Variable-temperature proton T_1 relaxation times, measured in a toluene-d_8 solution of complex $RuH_2(H_2)(PPh_3)_2$: circles hydride ligands; triangles para and meta protons of PPh_3; squares ortho protons of PPh_3. (Reproduced from D. G. Gusev, A. B. Vymenits, V. I. Bakhmutov. *Inorganica Chimica Acta* 1991; **179**: 195, with permission from Elsevier)

Figure 6.10, measured for all the protons in a toluene-d_8 solution of complex $RuH_2(H_2)(PPh_3)_2$ [8], do not correspond to Bloembergen–Purcell–Pounds' relaxation, the slopes of the curves in the high- and low-temperature zones are not identical and thus the Cole–Davidson correlation time distribution could be a good approximation for treatment of these data.

Finally, it should be emphasized that deviations of nuclear relaxation from of the isotropic motional model are well documented in solids, highly viscous liquids and glassy systems [1]. In contrast, a model of motion in regular liquids or solutions is still not clearly established. For this reason, we use the concept of the correlation time distribution as a mathematical tool for estimation of errors in the calculated internuclear distances.

Bibliography

1. P. A. Beckmann. *Physics, Reports* 1988; **171**: 85.

2. P. J. Desrosiers, L. Cai, Z. Lin, R. Richards, J. Halpern. *Journal of the America Chemical Society* 1991; **113**: 4173.

3. D. E. Woessner. *Journal of Chemical Physics* 1962; **36**: 1.

4. R. K. Harris. *Nuclear Magnetic Resonante Spectroscopy. A Physicochemical View.* Longman: London 1986.

5. D. G. Gusev, D. Nietlispach, A. B. Vymenits, V. I. Bakhmutov, H. Berke. *Inorganic Chemistry* 1993; **32**: 3270.

6. V. A. Daragan, M. A. Loczewiak, K. H. Mayo. *Biochemistry* 1993; **32**: 10580.

7. V. I. Bakhmutov, E. V. Vorontsov. *Reviews of Inorganic Chemistry* 1998; **18**: 183.

8. D. G. Gusev, A. B. Vymenits, V. I. Bakhmutov. *Inorganica Chimica Acta* 1991; **179**: 195.

7 ^{1}H T_1 Relaxation Diagnostics in Solution

Chapters 7–12 introduce practical applications of nuclear relaxation in solution, and focus on methodological aspects in studies of: weak intermolecular interactions (Chapter 7); complexations and associations (Chapter 11); chemical exchange in simple (Chapter 7) and complex (Chapter 11) molecular systems, internuclear distances, nuclear quadrupole coupling constants and chemical shift anisotropies (Chapters 8, 9 and 11); and paramagnetic molecular systems (Chapter 12).

The main chemical objects used to illustrate structural studies and their solutions on the basis of nuclear relaxation are transition metal hydride complexes. The choice of these complexes, as model systems, is subjective. However, it is quire reasonable from the chemical point of view for two reasons. First, the hydride systems show a large variety of bonding modes from normal (covalent) metal–hydrogen bonds in classical hydrides to dihydrogen complexes (or nonclassical hydrides) with η^2-bonded H_2 (Figure 7.1) [1]. Thus, determinations of H–H distances in such systems are very important for their structural formulation. Secondly, dihydrogen (η^2-H_2) ligands in nonclassical transition metal hydrides are very mobile and undergo various ultra-fast internal reorientations, even in the solid state. In other words, their hydride atoms are spatially unfixed and therefore they are excellent models to study the principles of nuclear relaxation in mobile groups (Chapter 10).

Chapter 7 considers applications of T_1 times of protons and deuterons for qualitative testing in solutions. Note that similar approaches can be used in studies of other nuclei.

Practical NMR Relaxation for Chemists   Vladimir I. Bakhmutov
© 2004 John Wiley & Sons, Ltd   ISBNs: 0-470-09445-1 (HB); 0-470-09446-X (PB)

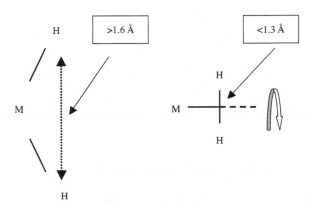

Figure 7.1 Classical (left) and non-classical (right) transition metal hydride systems with r(H–H) distances of >1.6 and <1.3 Å, respectively [1]. H–H bonding between hydrogen atoms in the non-classic systems leads to the formation of (η^2-H_2) ligands, capable of ultra-fast long-amplitude internal motions

7.1 Revealing Weak Intermolecular Interactions by T_1 Time Measurements in Solution

Interactions according to the equation:

$$A + B \rightleftharpoons A\cdots B \tag{7.1}$$

lead to formation of molecular aggregates, the energies of which depend on the nature of the interactions and on the solvent. As a rule, noncovalent intermolecular interactions such as electrostatic attraction, π–π interactions, specific solvatation and hydrogen bonding, are quite weak. For this reason, the formation of adducts (Equation 7.1) is fast on the NMR time scale and therefore the observed NMR signals and their parameters are averaged between free and bound states. For example, if the A molecules are detected, then:

$$\delta_A(obs) = P(A)\delta(A) + P(A\cdots B)\delta(A\cdots B)$$

$$1/T_{1A}(obs) = P(A)1/T_1(A) + P(A\cdots B)1/T_1(A\cdots B) \tag{7.2}$$

$$P(A\cdots B) = 1 - P(A)$$

where $P(A)$ and $P(A\cdots B)$ are mole fractions of the A and A$\cdots$B states, respectively [2].

Owing to the larger sizes and bigger moments of inertia of aggregates, their molecular motions become slower and correlation times τ_C increase. Since under conditions of fast molecular tumbling, relaxation rates $1/T_1$ are proportional to τ_C (with the exception of the spin–rotation mechanism, see

Equation 5.8), the formation of molecular aggregations will be accompanied by *decreasing* T_1 *times* measured for nuclei in the A (or B) states. It is easy to formulate a best condition for the T_1 diagnostics of weak interactions: a relaxation measurement is more sensitive to molecular aggregations when it is carried out on molecules of smaller size. This situation occurs, for example, in solutions, containing acidic and basic components ($(CF_3)_2CHOH$) and $(Bu_4N)_2[B_{12}H_{12}]$, respectively, which can interact to give $BH^{\delta-}\cdots^{\delta+}H$ dihydrogen bonds [3]. The geometry of one of the $H^{\delta-}\cdots^{\delta+}H$ complexes, optimized by the DFT method (in the gas phase) is shown in Figure 7.2. In solution, because of dihydrogen bonding with the bulky $(Bu_4N)_2[B_{12}H_{12}]$, an effective molecular inertia moment of the alcohol $(CF_3)_2CHOH$ will increase. This simple idea agrees well with the relaxation data in Table 7.1: the T_1 times, measured for ^1H and ^{19}F nuclei of the alcohol, reduce markedly upon addition of $(Bu_4N)_2[B_{12}H_{12}]$. In contrast, the effect, observed for ^{11}B nuclei in the bulky $(Bu_4N)_2[B_{12}H_{12}]$, molecules is smaller. In the presence of the alcohol, the ^{11}B T_1 time changes insignificantly from 0.0215 to 0.0179 s. Finally, it must be emphasized that the T_1 measurements have been made in dilute solutions to avoid the viscosity effects at the addition of $(Bu_4N)_2[B_{12}H_{12}]$ to a solution of $(CF_3)_2CHOH$ or vice versa.

The same idea can be used to reveal any noncovalent interactions. For example, the behavior of monoaromatic molecules (benzene and phenol etc.) in aqueous solutions, important for environmental chemistry, has been

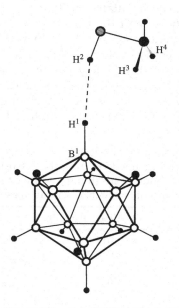

Figure 7.2 Geometry, optimized by the DFT method (gas phase) for the H...H bonded complex formed between the $[B_{12}H_{12}]^{2-}$ anion and methanol

Table 7.1　Room-temperature T_1 relaxation times (s) in CD_2Cl_2 solutions of individual compounds $(Bu_4N)_2[B_{12}H_{12}]$ and $(CF_3)_2$ CHOH and their mixture at a $(CF_3)_2CHOH/(Bu_4N)_2[B_{12}H_{12}]$ ratio of 1:2

$[B_{12}H_{12}]^{2-}$	$(CF_3)_2CHOH$	$[B_{12}H_{12}]^{2-} + (CF_3)_2CHOH$
^{11}B: 0.0215	^{19}F: 6.3	^{11}B: 0.0179 ^{19}F: 3.1
	^1HO: 5.4	^1HO: 1.9
	^1HC: 7.5	^1HC: 1.9

investigated by deuterium T_1 relaxation experiments in the presence of dissolved humic acids [4]. Since the deuterium quadrupole coupling constant of benzene-d_6 is well known (196.5 ± 1.3 kHz), the molecular motion correlation times τ_C can be calculated through $^2H\ T_1$ times (see Equation 4.24). On the other hand, the τ_C values can be obtained via the equation:

$$\tau_C = 4\pi f a^3 \eta / 3\ kT \tag{7.3}$$

if the viscosity η is measured experimentally. The microviscosity factor f in Equation (7.3), equal to $1/6$, accounts for the fact that diffusing molecules do not experience a continuous medium. Table 7.2 lists the viscosity, measured in benzene-d_6/water solutions, the τ_C values obtained via Equation (7.3) and the $^2H\ T_1$ times expected on the basis of these parameters. Figure 7.3 shows $^2H\ T_1$ times measured experimentally at different pH values.

As can be seen, in the case of the benzene dissolved in pure water, the calculated and experimental $^2H\ T_1$ times are very similar, and thus, in pure water, solvent viscosity is the primary factor, affecting the benzene molecular motions. The situation changes in the presence of humic acids. As seen, the measured $^2H\ T_1$ values are much smaller than those obtained in pure water. Thus, benzene molecules are interacting with the humic acids. Water solutions of phenol show the strongly reduced $^2H\ T_1$ times (0.1–0.25 s)

Table 7.2　T_1 relaxation times and molecular motion correlation times τ_C calculated for benzene molecules via Equation (7.3) from the viscosity data obtained in pure water and in the presence of humic acids (humic or fulvic acid concentration = 1 mg/ ml; $T = 295$ K; the molecular radius of benzene is taken as $2.49\ 10^{-10}$ m)

Solvent	Viscosity (cP)	τ_C (s)	T_1 (s)
Pure water	1.018	2.707×10^{-12}	0.646
Soil humic acid	1.010	2.685×10^{-12}	0.652
Peat humic acid	0.993	2.639×10^{-12}	0.663
Suwannee River humic acid	1.037	2.757×10^{-12}	0.635
Suwannee River fulvic acid	0.996	2.648×10^{-12}	0.661

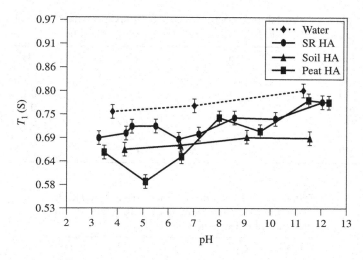

Figure 7.3 ^2H T_1 times, measured as a function of PH, for benzene-d$_6$ molecules in pure water and in the presence of soil humic acid, Suwannee River (SR) humic acid and peat humic acid. (Reproduced with permission from M. A. Nanny, J. P. Maza. *Environmental Science and Technology* 2001; **35**: 379. © 2001 American Chemical Society)

compared with 0.646 s (see Table 7.2) calculated for benzene. It is obvious that slower motions of the phenol are explained by the hydrogen bond formation between phenol and water molecules.

Because of the small moments of inertia of regular organic molecules and insignificant activation energies of their molecular reorientations in dilute (nonviscous) solutions, they do not show T_1 minima in variable-temperature experiments. However if one of the interacting components is bulky, then, because to the interaction, small molecules can also show T_{1min} times. In this context, it is interesting to consider ^1H relaxation of alcohol (CF$_3$)$_2$CHOH, dissolved in toluene-d$_8$ in the presence of Nb trihydride Cp$_2$NbH$_2^B$H^A. The latter could form dihydrogen-bonded species Cp$_2$NbH$_2^B$H$^A\cdots$HOCH(CF$_3$)$_2$ and Cp2NbH^{A}H$^B\cdots$HOCH(CF$_3$)$_2$ according to quantum chemical (DFT) calculations [5]. The complex Cp$_2$NbH$_2^B$H^A is quite bulky and, on cooling, ^1H T_1 relaxation times of the Cp, H^B and H^A resonances go through minima (Table 7.3). Table 7.4 lists the relaxation data, collected in the presence of a two-fold excess of (CF$_3$)$_2$CH−OH. Again, the ^1H T_1 times of Cp$_2$NbH$_2^B$H^A are minimal at low temperatures. However in addition, the CH resonance of the alcohol (CF$_3$)$_2$CH−OH also shows a minimal T_1 time (^1H $T_{1min} = 0.136$ s) observed at 210 K. This is direct experimental evidence for the H$\cdots$H aggregation occurring in solution.

Table 7.3 Variable-temperature ^1H T_1 relaxation data (s), collected for $Cp_2NbH_2{}^BH^A$ in pure toluene-d_8 at 400 MHz

T (K)	Cp	H^A	H^B
180	1.024	0.131	0.148
190	1.150	0.109	0.133
200	1.4	0.116	0.146
210	1.98	0.167	0.221
220	2.8	0.253	0.344
230	3.8	0.367	0.500

Table 7.4 ^1H T_1 times (s) measured at 400 MHz in a toluene-d_8 solution of $Cp_2NbH_2{}^BH^A$ in the presence of $(CF_3)_2CH-OH$ (1:1 ratio)

T (K)	Cp	CH	H^A	H^B
190	0.801	0.192	0.121	0.120
200	0.788	0.156	0.106	0.106
210	1.07	0.136	0.120	0.128
220	1.4	0.140	0.127	0.160
230	1.5	0.154	0.150	0.232
240		0.189	0.239	0.360

The same methodology can be applied for studies of H_2 binding to the iridium center:

$$IrHX_2(PR_3)_2 + H_2 \Longleftrightarrow trans\text{-}IrH(H_2)X_2(PR_3)_2 \tag{7.4}$$

$$X = Cl, Br; R = Pr^i, Cy$$

This reaction is fast and reversible on the ^1H NMR time scale [6, 7]. However, the principal feature of this system is the fact that the (H_2) ligand in the dihydrogen complex is 'spectrally unobservable', even at lowest temperatures, when the equilibrium is completely shifted to the right-hand side. In a pure toluene solution, the ^1H resonance of free H_2 appears as a narrow singlet (4.54 ppm) relaxing by a complicated mechanism. The ^1H T_1 time of free H_2 molecules is relatively long and depends slightly on the temperature, as shown in Figure 7.4. However, in the presence of the iridium hydride, the $T_1(H_2)$ time is greatly reduced, and the variable-temperature ^1H T_1 experiments reveal a T_1 minimum for this small molecule. This effect is good evidence for (H_2) binding to the Ir complexes. Moreover the ^1H T_1 times, averaged between the free and bound dihydrogen molecules:

$$1/T_1(H_2)^{OBS} = P(H_2)/T_1(H_2) + P(\eta^2\text{-}H_2)/T_1(\eta^2\text{-}H_2) \tag{7.5}$$

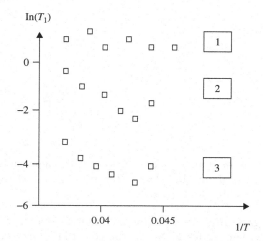

Figure 7.4 Variable-temperature ^1H T_1 data (200 MHz) for the signal of H_2 in toluene-d_8 solution in the absence (1) and in the presence of $IrHCl_2(PCy_3)_2$ (2); line 3 characterizes the T_1 times of the 'unobserved' (H_2) ligand in $IrH(H_2)Cl_2(PCy_3)_2$

allow one to calculate a ^1H T_{1min} time (3.4 ms at 200 MHz) for the 'spectrally unobservable' ($^2\eta$-H_2) ligand (see curve 3 in Figure 7.4).

7.2 T_1 Studies of Exchange in Simple Molecular Systems

The term 'simple molecular systems' means that such systems give the relatively simple NMR spectra, where signals are well resolved and provide direct measurements of relaxation times. Approaches to studies of exchange in complex molecular systems are considered in Chapter 11. As is known, kinetics and energetic parameters of chemical exchanges, occurring on the NMR time scale, can be obtained by a complete analysis of lineshapes [2] undergoing a typical temperature evolution, illustrated in Figure 3.6. Applications of the approximate formulas, based on an analysis of linewidths, are also possible [8]. However if exchanges are slow and resonance lines remain narrow, showing practically natural linewidths, these approaches are not effective. In this situation, exchange can be characterized by the relaxation technique. Actually, as we have shown in Chapter 3, the presence of slow exchange is easy revealed by saturation transfer experiments. These experiments, in combination with the T_1 time measurements give kinetic parameters of revealed exchanges.

Hydride ligands of ruthenium complexes 1–4 in Figure 7.5 show two narrow ^1H resonances in a large temperature region. However, on H(X)

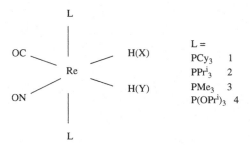

Figure 7.5 The rhenium dihydride complexes with magnetically non-equivalent hydride ligands undergoing slow H(X)/H(Y) positional exchanges in the variable-temperature ^{1}H NMR spectra (Table 7.5)

irradiation, integral intensities of the H(Y) lines decrease markedly. The same effects are observed for H(X) resonances at H(Y) irradiations [9]. Thus, a slow hydride/hydride exchange occurs in these systems and its rate constant k can be determined via the equation:

$$k = (1/T_1(\text{Y}))\{(I^0/I^{ST}) - 1\} \tag{7.6}$$

where I^{ST} and I^0 are the H(Y) integral intensities in the presence and absence of H(X) irradiations, respectively. Note that the $T_1(\text{Y})$ time in Equation (7.6) is measured in the presence of the saturating radiofrequency field. In the case of complexes 1–4, T_1 times of the H(Y) and H(X) ligands are practically identical, but they differ widely within this series of compounds: the smallest value is observed for complex 1 (0.26 s at 16°C) and the largest value is found for complex 2 (3.3 s at 16°C). Finally, the variable-temperature saturation transfer and T_1 experiments give the exchange rates and the activation parameters in Table 7.5.

The pulse sequence:

$$\text{RD} - 180^\circ_{\text{SEL}} - \text{t} - 90^\circ - \text{AT} \tag{7.7}$$

Table 7.5 Kinetic parameters of H(X) / H(Y) exchange in complexes 1–4 (Figure 7.5) obtained by saturation transfer experiments on the hydride resonances $(\tau(\text{H}(X)) = \tau(\text{H}(Y)) = 1/k)$

Complex	Solvent	τ (s) /T(C)	$\Delta H^{\neq}$ (kcal/mol)	$\Delta S^{\neq}$ (e. u.)
1	Toluene-d_8	1.2/60	18.4	−3.7
1	Toluene-d_8/DMSO-d_6 (1:1)	2.4/60	18.7	−3.9
2	Toluene-d_8	0.95/60	20.5	2.3
3	Toluene-d_8	0.57/60	15.2	−12.4
4	Toluene-d_8	0.055/60	14.6	−9.1
		0.15/45		
4	Toluene-d_8/DMSO-d_6 (1:1)	0.2/45	13.6	−13.3

applied for measurements of the selective T_1 times, can also lead to determinations of lifetimes for the slowly exchanging A and X resonances. Note that the first 180° pulse in sequence (7.7) selectively excites, for example, A nuclei and the delay time t varies. If the $T_1(A)$ and $T_1(X)$ relaxation times are identical, then evolution of integral intensities $I(t)$ as a function of the t time in the equation:

$$[\{I_0A - IA(t)\} - \{I_0X - IX(t)\}]/[\{I_0A - IA(t)\} - \{I_0X - IX(t)\}] = \exp(-2t/\tau_A)$$
(7.8)

provides calculations of the τ_A lifetimes [8]. However, this method is effective when the determined lifetimes are shorter or comparable to the $T_1(A)$ and $T_1(X)$ times. In other words, the A and X resonances should have relatively long relaxation times.

The radiofrequency field B_1, operating in a spin-locking experiment, is weaker by several orders of magnitude than the external magnetic field B_0. In this case, frequencies of molecular motions, governing the $T_{1\rho}$ relaxation lie in a diapason of tens of kHz. If chemical exchanges occur with frequencies of the same order, they can also affect $T_{1\rho}$ times, but not T_1 times. Thus, in the presence of exchange, the $T_{1\rho}$ and T_1 times are not equal. Note that exchanges with frequencies of tens of kHz completely average the A and X resonances. For this reason, ratios $T_{1\rho} < T_1$, obtained for averaged resonances, are a good test for the presence of fast chemical exchanges. By definition, exchange contributions can be expressed as:

$$T_1(EXCH)^{-1} = (T_{1\rho})^{-1} - (T_1)^{-1}$$
(7.9)

Then lifetimes $\tau(\tau = \tau_A = \tau_X)$ can be obtained by experiments at various powers of the spin-locking field $(\omega_1 = \gamma B_1)$ via the equation:

$$T_1(EXCH) = \{(1/2)\pi^2(\nu^A - \nu^X)^2\tau\}^{-1} + \tau\omega_1^2\{2\pi^2(\nu^A - \nu^X)^2\}^{-1}$$
(7.10)

Additionally, these measurements, represented as plots of the $T_1(EXCH)$ versus ω_1^2, give the τ values and simultaneously the chemical shift differences, $(\nu^A - \nu^X)$.

7.3 Structural ^1H T_1 Criterion

The rates of dipole–dipole nuclear relaxation are inversely proportional to the sixth power of the internuclear distances. It is obvious that, for similar molecular mobility within a series of chemical compounds, shorter internuclear distances will cause shorter relaxation times. Practical application of this principle is particularly effective for structural formulation of systems with very short H...H distances [10]. This situation occurs in dihydrogen complexes where $r(H–H) < 1$ Å (see Figure 7.1). Table 7.6 lists the ^1H T_1 times,

Table 7.6 ^{1}H T_1 times measured by inversion recovery experiments at 250 MHz for the hydride resonances in solutions of classical and nonclassical transition metal hydride complexes

Hydride	T_1 (ms)	Conditions
H_2	1600	Toluene-d_8, 203 K
$IrH_5(PCy_3)_2$	820	CD_2Cl_2, 193 K
$H_2Fe(CO)_4$	3000	Toluene-d_8, 203 K
$ReH_5(PPh_3)_3$	540	Toluene-d_8, 203 K
$OsH_4(PTolyl)_3)_3$	820	Toluene-d_8, 203 K
$ReH_8(PPh_3)_3^-$	245 ($T_{1\,min}$)	Ethanol-d_6, 200 K
$WH_5(PMePh_2)_4^+$	179 ($T_{1\,min}$)	CD_2Cl_2, 240 K
$WH_6(PMe_2Ph)_3$	181 ($T_{1\,min}$)	Toluene-d_8, 235 K
$W(H_2)(CO)_3(PPr^i_3)_2$	5 ($T_{1\,min}$)	Toluene-d_8
$Ru(H_2)H_2(PPh_3)_3$	38	Toluene-d_8, 203 K
$RuH(H_2)(CO)(triphos)$	6.2/51	CD_2Cl_2, 190 K
$trans$-$IrH(H_2)Cl_2(PCy_3)_2$	4.2 ($T_{1\,min}$)	Toluene-d_8, 235 K
$Os(H_2)H_3(PPh_3)_3^+$	35 ($T_{1\,min}$)	CD_2Cl_2, 220 K
$[IrH(H_2)(bq)(PPh_3)_2]^+$	8 ($T_{1\,min}$)	CD_2Cl_2, 200 K

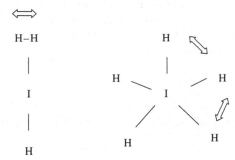

Figure 7.6 Proton–proton dipole–dipole interactions (see corresponding arrows) in the *trans* dihydrogen/hydride iridium complex and the classical iridium polyhydride. The scheme considers only shortest hydride/hydride contacts

measured in solutions of transition metal hydride complexes at 250 MHz. Similar temperatures minimize the effects of the molecular motion correlation times τ_C. The (H$_2$) ligand in the dihydrogen complex IrH(H$_2$)Cl$_2$(PCy$_3$)$_2$ (Figure 7.6) is located *trans* to the closest hydride atom while each hydride ligand in the classical complex IrH$_5$(PCy$_3$)$_2$ is adjacent to two hydride atoms. Nevertheless, the (H$_2$) resonance in the dihydrogen complex shows a very short T_1 time of 4.2 ms compared with 820 ms observed in IrH$_5$(PCy$_3$)$_2$. The classical dihydride H$_2$Fe(CO)$_4$ also exhibits an extremely long relaxation

Table 7.7 ^1H $T_{1\,min}$ values calculated at 250 MHz for: (a) slow- and fast-spinning dihydrogen ligands (SS and FS, respectively) at H−H distances 0.75−1.35 Å; (b) two terminal hydride ligands separated by 1.65 Å

$r(H−H)$ Å	$T_{1\,min}$ (ms)	
	SS	FS
0.75	1.1	4.4
1.0	6.5	26
1.1	11.5	45.8
1.2	19.3	77.2
1.3	31.2	125
1.35	39	157
1.65	130	

time. The $T_1(H_2)$ time in dihydrogen complex $W(H_2)(CO)_3(PPr^i_3)_2$ is again short. Thus, comparison of these data leads to an empirical formulation of the T_1 criterion at 250 MHz: $T_1 > 150$ ms for classical hydrides and $T_1 < 80$ ms for dihydrogen complexes.

To verify the reliability of such a formulation, consider a pair of ^1H nuclei in the so-called elongated dihydrogen complexes where H−H distances can reach 1.35 Å [1]. Calculations of ^1H $T_{1\,min}$ values allow one to avoid the influence of τ_C on T_1. Under these conditions Equation (4.9) coverts to:

$$T_{1\,min}(H−H) = \{(r(H−H)/5.815\}^6\,\nu \qquad (7.11)$$

where $T_{1\,min}$, $r(H−H)$ and ν are measured in seconds, Å and MHz, respectively [11]. Now calculate the ^1H $T_{1\,min}$ time at 250 MHz as a function of the $r(H−H)$ for immobile and rapidly spinning (see Figure 7.1) dihydrogen ligands. Remind that fast (H_2) rotation causes a four-fold ^1H T_1 increase (Equation 6.7). Table 7.7 shows the data obtained where a classical dihydride structure with a H−H separation of 1.65 Å is taken for comparison. It is obvious that the elongated and rapidly spinning dihydrogen ligands and the classical hydride atoms are not distinguishable. On the other hand, even rapidly spinning dihydrogen ligands with short H−H distances ($\leqslant 1$ Å) can be reliably identified by ^1H $T_{1\,min}$ measurements in solution. It is easy to show that the $T_{1\,min}$ criterion is still valid when the dihydrogen ligands are involved in a fast chemical exchange with terminal hydride atoms (Figure 7.7).

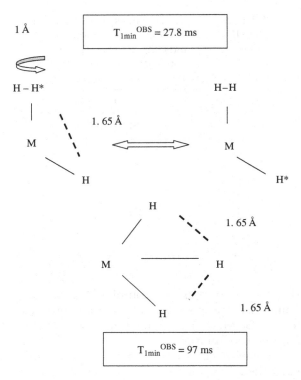

Figure 7.7 ^{1}H $T_{1\,min}$ times calculated at 250 MHz for: a resonance of a hydride/dihydrogen complex with a fast-spinning (H$_2$) ligand and fast hydride/dihydrogen exchange; a resonance of a classical trihydride system

7.4 Partially Relaxed NMR Spectra

Partially relaxed NMR spectra, recorded by standard inversion recovery experiments with time delays $\tau < T_1$, are also useful for structural diagnostics in solution. For example, the room-temperature reaction:

$$ReH_2(CO)(NO)(PR_3)_2 + (CH_3)_2C{=}O + CF_3COOH \overset{-70^\circ C}{\Rightarrow}$$

$$[ReH(CO)(NO)(PR_3)_2\{(CH_3)_2CH - OH\}]^+[CF_3COO]^{-}\overset{25^\circ C}{\Rightarrow}$$

$$[ReH(CO)(NO)(CF_3COO)(PR_3)_2] + (CH_3)_2CHOH \qquad (7.12)$$

yields the Re monohydride and the alcohol, as products of an ionic hydrogenation. These products can be easy identified by the traditional ^{1}H NMR spectra. However, a low-temperature reaction (7.12) leads to an intermediate compound where the alcohol molecule can bind to the metal center [12]. Figure 7.8 shows schematically the partially relaxed ^{1}H NMR spectrum

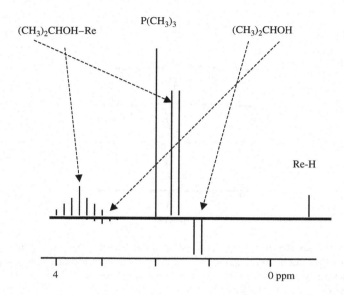

$(CH_3)_2CHOH-Re$ $\quad$ $P(CH_3)_3$ $\quad$ $(CH_3)_2CHOH$

Re-H

4 $\qquad$ 0 ppm

Figure 7.8 Partially relaxed ^{1}H NMR spectrum (schematic), recorded at $-70°C$ in a toluene solution after the low-temperature hydrogenation reaction (Equation 7.12). The spectrum is obtained by the inversion recovery experiment at $\tau < 0.5$ s. The signals with positive and negative phases belong to the alcohol complex $[ReH(CO)(NO)(P((CH_3)_3)_2\{(CH_3)_2CHOH\}]^+$ $[CF_3COO]^-$ and free $(CH_3)_2CHOH$

recorded at the low temperature. As can be seen, the resonances of the organometallic and alcohol fragments are practically relaxed while a free alcohol molecule (the lines with negative phases) is far from full relaxation. In other words, motions of the organometallic and coordinated alcohol moieties are correlated, and thus the partially relaxed spectra strongly support the structure of the low-temperature intermediate.

Owing to small chemical shift differences, the signals observed in NMR spectra often overlap to give broadened lines. Partially relaxed spectra can help in studies of such lines when T_1 times of the overlapped resonances are different. For example, after shaking a toluene-d_8 solution of $RuH_4(PPh_3)_3$ under a D_2 atmosphere, the ^{1}H NMR spectra reveal a D_2/H_2 exchange:

$$RuH_4(PPh_3)_3 + D_2 \Longleftrightarrow RuDH_3(PPh_3)_3 + HD \qquad (7.13)$$

$$Ru\,DH_3(PPh_3)_3 + HD \Longleftrightarrow RuD_2H_2(PPh_3)_3 + H_2$$

In fact, the initial integral intensity of the hydride resonance in $RuH_4(PPh_3)_3$ (-7.2 ppm) decreases and typical signals of H_2 and HD molecules appear at ~ 4.5 ppm. The residual hydride line is strongly broadened (30 Hz) and thus the resonances of isotopomers are not resolved. In contrast, the partially relaxed ^{1}H NMR spectra (Figure 7.9) [13] recorded by inversion recovery

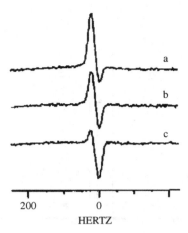

Figure 7.9 ^{1}H NMR signals of hydride ligands of the incompletely deuterated complex $RuH_4(PPh_3)_3$ collected by the inversion recovery method at 280 K. (Reproduced from D. G. Gusev, A. B. Vymenits, V. I. Bakhmutov. *Inorganic Chimica Acta* 1991; **179**: 195, with permission from Elsevier)

experiments with variations in the τ time, clearly show that the broadened line consists of two resonances with chemical shifts of -7.21 and -7.11 ppm. The former resonance, showing a smaller relaxation rate, can be assigned to $RuHD_3(PPh_3)_3$. The second line corresponds to $RuH_2D_2(PPh_3)_3$. It must be emphasized that this methodology is quite suitable for any nuclei.

Bibliography

1. G. J. Kubas. *Metal Dihydrogen and σ-Bond Complexes*. Kluwer/Plenum: New York, 2001.

2. J. H. Kaplan, G. Frenkel. *NMR of Chemically Exchanging Systems*; Academic Press: New York, 1980.

3. E. S. Shubina, E. V. Bakhmutova, A. M. Filin, I. B. Silaev, L. N. Teplitskaya, A. L. Chistyakov, I. V. Stankevich, V. I. Bakhmutov, V. I. Bregadze, L. M. Epstein. *Journal of Organometallic Chemistry* 2002; **657**: 155.

4. M. A. Nanny, J. P. Maza. *Environmental Science and Technology* 2001; **35**: 379.

5. E. V. Bakhmutova, V. I. Bakhmutov, G. Nikonov, A. LLedos *et al.* *Chemistry A European Journal* 2004; **10**: 661.

6. V. I. Bakhmutov, E. V. Vorontsov. *Reviews of Inorganic Chemistry* 1998; **18**: 183.

7. V. I. Bakhmutov, E. V. Vorontsov, A. B. Vymenits. *Inorganic Chemistry* 1995; **34**: 214.

8. R. K. Harris. *Nuclear Magnetic Resonance Spectroscopy*. Longman: London, 1986.

9. V. I. Bakhmutov, T. Burgi, P. Burger, H. Berke. *Organometallics* 1994; **13**: 4203.

10. D. G. Hamilton, R. H. Crabtree. *Journal of the American Chemical Society* 1988; **110**: 4126.

11. P. J. Desrosiers, L. Cai, Z. Lin, R. Richards, J. Halpern. *Journal of the American Chemical Society* 1991; **113**: 4173.

12. V. I. Bakhmutov, E. V. Vorontsov, D. Y. Antonov. *Inorganica Chimica Acta* 1998; **278**: 122.

13. D. G. Gusev, A. B. Vymenits, V. I. Bakhmutov. *Inorgica Chimica Acta* 1991; **179**: 195.

8 Internuclear Distances from ^{1}H T_1 Relaxation Measurements in Solution

If relaxation in a pair of nuclei is monoexponential and completely governed by hetero- or homonuclear dipolar coupling in this pair, then measurements of relaxation rates is a direct way to determine interatomic distances. However, in most cases, nuclei relax by a combination of different mechanisms or different dipolar interactions. Such a situation is typical of molecular systems where, for example, X−H or H−H distances are determined by NMR experiments on target nuclei (^{1}H, ^{13}C, ^{19}F or ^{31}P), which are surrounded by a great number of protons. In such cases, a necessary dipolar contribution must be correctly evaluated from a measured (total) relaxation rate. In this chapter, we consider the methodology of approaches to separations of dipolar contributions based on the direct ^{1}H T_1, ^{1}H T_{1sel} and ^{1}H T_{1bis} measurements or their combinations and also on so-called indirect relaxation experiments. Finally, we analyze errors in determinations of internuclear distances.

Practical NMR Relaxation for Chemists   Vladimir I. Bakhmutov
© 2004 John Wiley & Sons, Ltd   ISBNs: 0-470-09445-1 (HB); 0-470-09446-X (PB)

8.1   X–H Distances: Metal–Hydride Bond Lengths

As we have shown, the strength of heteronuclear dipolar coupling depends on the nature of interacting nuclei. Since protons have the largest γ value and practically 100% natural abundance (Table 1.1), dipole–dipole interactions involving their participation are most significant. At a constant internuclear distance in an H–X pair, the heteronuclear dipolar coupling reduces proportionally to coefficient k, calculated via the equation:

$$k = 0.44 \, (v_X/v_H)^2 I_X(I_X + 1) \tag{8.1}$$

where v_H and v_X are the NMR frequencies of H and X nuclei, respectively, and I_X is the spin of X nuclei. Table 8.1 lists the k values, obtained by Equation (8.1) together with the natural abundance of X nuclei. It is easy to see that determinations of internuclear distances or bond lengths will be problematic for ^1H–^{14}N or ^1H–^{103}Rh pairs because of the weak dipolar coupling, particularly when a large number of protons surrounds the target nuclei. Proton–boron dipole–dipole interactions are considerably stronger and, for example, the ^1H and ^{11}B T_1 measurements in $NaBH_4$ lead to a B–H distance of 1.26 Å, in good agreement with the neutron diffraction data [1]. The natural abundance of ^{13}C and ^{15}N isotopes is too low and therefore ^1H T_1 experiments are not appropriate for determinations of H–C and H–N bond lengths. On the other hand, these distances are determinable by T_1 experiments on ^{13}C or ^{15}N nuclei. For example, ^{15}N T_1 relaxation studies of labeled ^{13}C–^{15}N fragments finally lead to C–N distances when the ^{13}C–^{15}N dipolar contributions are precisely evaluated from relaxation

Table 8.1 Strength of dipolar coupling in pairs of H/X nuclei as a function of the nature of X nuclei in the external magnetic field of 7.05 T

Isotope	Spin I	v_0 (MHz)	k	Natural abundance (%)
^1H	1/2	300	1	100
^{11}B	3/2	96.2	0.14	80.1
^{31}P	1/2	121.4	0.054	100
^{14}N	1	21.7	0.0046	99.6
^{103}Rh	1/2	9.6	0.0003	100
^{55}Mn	5/2	74.3	0.24	100
^{59}Co	7/2	71.2	0.39	100
^{93}Nb	9/2	73.4	0.65	100
^{187}Re	5/2	68.2	0.13	62.6
^{181}Ta	7/2	36.0	0.1	99.9
^{51}V	7/2	78.9	0.48	99.7
^{195}Pt	1/2	64.5	0.015	33.8

rates, measured experimentally. The latter dictates the real accuracy of C—N determinations. In the solid state these dipolar contributions can be extracted by removing the effects of chemical shift anisotropy in the so-called DRAMA and REDOR experiments [2]. Then, the internuclear distances can be found by simulations of lineshapes in NMR spectra obtained as parts of two-dimensional experiments.

In spite of weak dipole—dipole interactions in the ^{1}H—^{195}Pt pair, this case is very convenient for T_1 relaxation studies in solution. In fact, a ^{1}H NMR spectrum of an H—Pt molecular fragment exhibits the central resonance and two satellite lines, corresponding to ^{1}H nuclei, attached to the nonmagnetic and magnetic Pt isotopes, respectively (Figure 8.1). It is obvious that, if relaxation times of satellite lines are shorter, then ^{1}H—^{195}Pt dipolar contributions are easy calculated as $1/T_1$(sat) − $1/T_1$(centr) [3]. A similar situation can be expected for mercury hydride complexes H^{199}HgR. In contrast, the central and satellite lines of the H^A ligand, binding to the W center in binuclear complex Cp$_2$TaH$^X_2$H^A—W(CO)$_5$, relax identically [4]. In fact, the NMR frequency of ^{183}W nuclei is too low (12.5 MHz) to provide a significant channel of nuclear relaxation.

It follows from Table 8.1 that the large spin numbers of ^{93}Nb, ^{51}V, ^{55}Mn, ^{187}Re, ^{59}Co and ^{181}Ta nuclei compensate their lower Larmor frequencies. For this reason, these H—M bond lengths can be successfully determined by ^{1}H T_1 relaxation measurements in solution. As we show below, the best way is location of minimal ^{1}H T_1 times by variable-temperature experiments. When ^{1}H relaxation is completely governed by proton—metal dipole—dipole interactions, metal—hydride distances r(M—H) are easy determined via

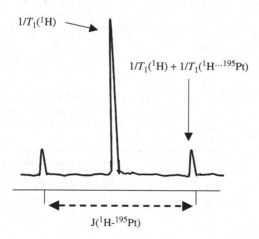

Figure 8.1 ^{1}H NMR spectrum and relaxation rates, $1/T_1$, of a Pt-hydride molecule (schematically)

Equation (4.4), rewritten in the form [5]:

$$r(\mathrm{M-H}) = C_{\mathrm{M-H}}(200\, T_{1\,\mathrm{min}}/\nu_{\mathrm{H}})^{1/6} \qquad (8.2)$$

where $r(\mathrm{M-H})$, $T_{1\,\mathrm{min}}$, and ν_{H} are measured in Å, seconds and MHz, respectively. In turn, coefficients $C_{\mathrm{M-H}}$ in Equation (8.2) can be calculated by using the magnetic properties in Tables 1.1 and 8.1. For example, Ta, Co, Nb, Re and Mn nuclei give $C_{\mathrm{M-H}}$ equal to 2.001, 2.491, 2.722, 2.228 and 2.287, respectively. It is easy to show that $C = 2.405$ for a pair of protons. Thus, the metal–hydride and proton–proton dipole–dipole interactions are comparable [5, 6].

These considerations show that significant metal–hydride dipole–dipole interactions shorten relaxation times of hydride resonances. For example, hydride ligands with identical environments in two structurally related complexes PP_3Rh-H and PP_3Co-H demonstrate the quite different ^1H $T_{1\,\mathrm{min}}$ times: 543 and 62 ms in PP_3Rh-H and PP_3Co-H, respectively (Table 8.2). It should be emphasized that the strong cobalt–hydride coupling in classical cobalt hydrides can lead to short relaxation times close to those measured in dihydrogen complexes (compare the $T_{1\,\mathrm{min}}$ values in complexes $[CoH_2(dppe)_2]^+$ and $ReCl(H_2)(dppe)_2$). It is obvious that such interactions can play a significant role, even in the ^1H T_1 behavior of dihydrogen ligands. Calculate, for example, ^1H $T_{1\,\mathrm{min}}$ times of ^1H relaxation by proton–proton and niobium–proton dipole–dipole interactions at 200 MHz in a quite reasonable model of dihydrogen complexes $Nb-(H_2)$ (Figure 8.2). If the (H_2) ligand is not spinning, then the ^1H $T_{1\,\mathrm{min}}(H-H)$ and ^1H $T_{1\,\mathrm{min}}(Nb-H)$ values are calculated via Equations (7.11) and (8.2) as 0.0052 and 0.0836 s, respectively. Thus, Nb–H dipolar interactions provide 5.9% of the total relaxation rate of an immobile dihydrogen ligand. According to Woessner's expression (see Equation 6.7), a fast (H_2) rotation partially averages the proton–proton dipolar coupling, leading to a four-fold ^1H $T_{1\,\mathrm{min}}$ elongation. By analogy, a (H_2) rotation rapidly reorients the Nb–H vector and thus niobium–proton dipole–dipole interactions reduce proportionally to a factor $(3\cos^2\theta - 1)^2/4$

Table 8.2 ^1H $T_{1\,\mathrm{min}}$ times measured in solutions of the classical transition metal hydride complexes with strong metal–hydride dipole–dipole interactions (recalculated for 250 MHz)

Compound	^1H $T_{1\,\mathrm{min}}$ (ms) Solvent
$CoH(dppe)_2$	50.0 Toluene-d_8
$[CoH_2(dppe)_2]^+$	53.7 Toluene-d_8
PP_3CoH	62.0 TDF
$[PP_3CoH_2]^+$	51.5 TDF
PP_3RhH	543.0 TDF
$ReCl(H_2)(dppe)_2$	53.7 Toluene-d_8

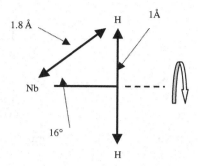

Figure 8.2 Proton−niobium and proton−proton dipole−dipole interactions in a niobium dihydrogen complex with a fast-spinning (H_2) ligand

where $\theta = 16°$ (Figure 8.2). Under these conditions, the ^1H $T_{1\,min}$(H−H) and ^1H $T_{1\,min}$(Nb−H) times are 0.021 and 0.107 s, respectively, and the Nb−H dipolar contribution gives 16.3% of the total ^1H relaxation rate. Such contributions cannot be ruled out in relaxation studies of dihydrogen complexes.

Finally, it must be emphasized that the metal−hydride dipole−dipole contributions will increase in binuclear hydride systems where H ligands are bridging between two metal atoms. For example, the rhenium−hydride dipolar coupling governs 90% of the ^1H T_1 relaxation rate in complex $(\mu\text{-}H)_2Re_2(CO)_8$.

8.1.1 How to Determine Metal−Hydride Bond Lengths by Standard ^1H T_1 Measurements

Dipolar coupling between protons and ^{14}N, ^{13}C or ^{17}O nuclei is too weak, and therefore hydride resonances in the complexes, depicted in Figure 8.3, relax completely by metal−hydride dipole−dipole interactions. For these reasons, ^1H $T_{1\,min}$ times, measured by standard inversion recovery experiments, directly give M−H bond lengths via Equation (8.2). For example, the Mn−H bond length in $MnH(CO)_5$, determined by the NMR relaxation technique in solution, is as long as 1.65 ± 0.05 Å [7] in good agreement with 1.60 ± 0.03 Å obtained by neutron diffraction in the solid state. Note that a slight elongation of the Mn−H distance, found in solution, can be connected with fast Mn−H vibrational motions (see Chapter 10).

Figure 8.4 illustrates a more typical situation, where the H ligand is adjacent to another hydride atom and in addition, to a number of protons and phosphorus nuclei located in the ligand environment. Under these conditions, the ^1H total relaxation rate, $1/T_1^{TOT}$, is:

$$1/T_1^{TOT} = 1/T_1(H-HM) + 1/T_1(M-H) + 1/T_1(H-H) + 1/T_1(H-P) + 1/T_1^*$$

$$(8.3)$$

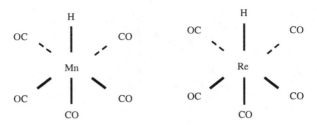

Figure 8.3 The octahedral Mn and Re hydride complexes, where metal–hydride dipole–dipole interactions completely control spin–lattice relaxation rates of hydride ligands

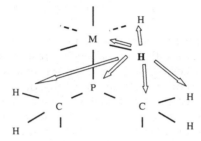

Figure 8.4 Schematic representation of hydride–hydride, hydride–metal, hydride–proton and hydride–phosphorus dipole–dipole interactions contributing to the ${}^1H$ spin–lattice relaxation of the marked hydride ligand

where $1/T_1(H-HM)$, $1/T_1(M-H)$, $1/T_1(H-H)$ and $1/T_1(H-P)$ correspond to relaxation contributions caused by hydride–hydride, hydride–metal, hydride–proton and hydride–phosphorus dipolar couplings, respectively. A contribution from other relaxation mechanisms (for example, the CSA mechanism) is shown as $1/T_1{}^*$. Since hydride–phosphorus dipolar interactions are relatively small (see the k values in Table 8.1), the protons, located in the ligand environment, make one of the principal contributions to the relaxation rate of hydride resonances. This contribution can be 'removed chemically' with a total deuteration of phosphorus ligands. Then, the hydride relaxation in deuterated molecules will be governed by metal–hydride interactions and ${}^1H\ T_1$ measurements will lead to determinations of the M–H bond lengths. This methodology has been used in studies of trihydride $(Cp)_2NbH^X{}_2H^A$ and its deuterated derivatives $(Cp)_2NbD^XH^XH^A$ and $(Cp)_2NbD^XD^XH^A$ [8] (Figure 8.5). Referring back to chapter 4 Figure 4.6 shows the ${}^1H$ NMR spectrum of the isotopomeric mixture in an H^A region. Owing to a significant deuterium perturbation of the chemical shift (the so-called isotopic chemical shift), all the H^A resonances are well resolved and the corresponding ${}^1H\ T_1$ times can be accurately measured. It follows from Figure 8.5 that the displacement of ${}^1H$ for ${}^2H$ is accompanied by an increase of relaxation

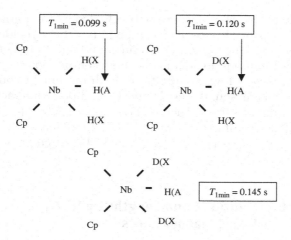

Figure 8.5 Structures of the niobium trihydride and its partially-deuterated derivatives, $(Cp)_2NbD(X)H(X)H(A)$ and $(Cp)_2NbD(X)_2H(A)$ [8]. The ^1H T_1 times are measured for the H(A) ligands in a toluene-d_8 solution at 400 MHz and 220 K

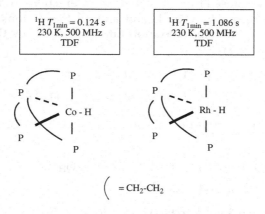

Figure 8.6 The structurally-similar cobalt and rhodium monohydrides, illustrating the contribution of metal–hydride dipole–dipole contributions to the total T_1 relaxation rates of hydride ligands

times. The longest ^1H T_1 time ($T_{1min} = 0.145$ s, 400 MHz) is measured for the complex in which the H^A ligand is surrounded by two deuterons. The niobium–hydride interactions are dominant in $(Cp)_2NbD^XD^XH^A$ and thus the Nb–H(A) distance can be calculated via Equation (8.2). As has been noted, the CSA relaxation is often responsible for the appearance of an additional contribution $1/T_1^*$ in Equation (8.3). In the case of the Nb trihydride, this mechanism is not effective because the experiments in a high-temperature zone at 200 and 400 MHz show identical ^1H T_1 times.

Metal–hydride contributions are well evaluated by comparison of nuclear relaxation in structurally similar compounds. The hydride systems in Figure 8.6 have the same ligand environment [9] and thus $1/T_1(H-H)$ and $1/T_1(H-P)$ contributions are identical in both compounds. Since the γ value of ^{103}Rh nuclei is very low, the rhodium–hydride dipolar coupling is negligible and for this reason, the H–H and H–P dipolar interactions, govern a minimal T_1 time of 1.086 s. Then, $1/T_{1\,min}^{TOT} = 1/1.086 + 1/T_{1\,min}(Co-H)$ and the $T_{1\,min}(Co-H)$ time is calculated as 0.140 s. Finally, this value gives, via Equation (8.2), a Co–H bond length of 1.54 Å in good agreement with the solid-state data [9].

8.1.2 Metal–Hydride Bond Lengths by $^1H\ T_{1sel}$ and $^1H\ T_{1\,min}$ Measurements

Metal–hydride contributions (as well as other heteronuclear dipolar contributions) can be evaluated spectroscopically by nonselective (T_1) and selective (T_{1sel}) relaxation experiments. Take $1/T_1(H-M)$ and $1/T_1(H-H)$ as relaxation rates caused by metal–hydride interactions and dipole–dipole interactions between a hydride ligand and all the protons in a molecule, respectively. Then the total relaxation rates, measured by the nonselective and selective pulses, are:

$$1/T_1 = 1/T_1(H-M) + 1/T_1(H-H) \tag{8.4}$$

$$1/T_{1sel} = 1/T_1(H-M) + 1/T_{1sel}(H-H) \tag{8.5}$$

The corresponding proton–proton and metal–proton contributions are written as:

$$1/T_1(H-H) = 0.3\,\gamma_H^4\,\hbar^2\,r(H-H)^{-6}\{\tau_c/(1+\omega_H^2\tau_c^2)$$
$$+ 4\tau_c/(1+4\omega_H^2\tau_c^2)\} \tag{8.6}$$

$$1/T_{1sel}(H-H) = 0.3\,\gamma_H^4\,\hbar^2\,r(H-H)^{-6}\{\tau_c/(1+\omega_H^2\tau_c^2)$$
$$+ 2\tau_c/(1+4\omega_H^2\tau_c^2) + \tau_c/3\} \tag{8.7}$$

$$1/T_1(M-H) = (2/15)\gamma_H^2\,\gamma_M^2\,\hbar^2\,I(I+1)r(M-H)^{-6} \times \{3\tau_c/(1+\omega_H^2\tau_c^2)$$
$$+ 6\tau_c/(1+(\omega_H+\omega_M)^2\tau_c^2) + \tau_c/(1+(\omega_H-\omega_M)^2\tau_c^2)\} \tag{8.8}$$

If T_1 and T_{1sel} times are measured in high-temperature zones, i.e. at $\omega_H^2\tau_c^2 \ll 1$, then the ratio between proton–proton and metal–proton contributions k can be expressed via:

$$k = \{0.3\,\gamma_H^4\,\hbar^2\,r(H-H)^{-6}\}/\{(2/15)\gamma_H^2\,\gamma_M^2\,\hbar^2\,I(I+1)r(M-H)^{-6}\}$$
$$= (f-1)/(0.5-f/3) \tag{8.9}$$

where $f = T_{1sel}/T_1$ [10]. Finally a combination of Equations (8.4), (8.5) and (8.9) gives:

$$1/T_1 = (4/30)r(M-H)^{-6}\, \gamma_H^2\, \gamma_M^2\, \hbar^2\, I(I+1)[(3+k) \times \tau_c/(1+\omega_H^2\tau_c^2)$$
$$+ 4k\tau_c/(1+4\omega_H^2\tau_c^2) + 6\tau_c/(1+(\omega_H+\omega_M)^2\tau_c^2)$$
$$+ \tau_c/(1+(\omega_H-\omega_M)^2\tau_c^2)] \tag{8.10}$$

Thus, if the T_{1sel} times are measured in a high-temperature region, the variable-temperature T_1 data can be fitted to Equation (8.10) to give the metal–hydride distance, $r(M-H)$, the activation energy of molecular reorientations E_a, and the correlation constant τ_0. Note that the fitting procedures can be performed with the least-squares linear regression programs in the commercial software Matlab package. Figure 8.7 illustrates fitting the relaxation data, collected in a toluene-d8 solution of complex $ReH_2(CO)(NO)(POPr^i_3)_2$. As can be seen, Equation (8.10) closely follows the experimental T_1 times in the large temperature diapason. Moreover, the E_a and τ_0 values, obtained by ^1H T_1 relaxation experiments at 200 and 300 MHz, are practically identical (Figure 8.8).

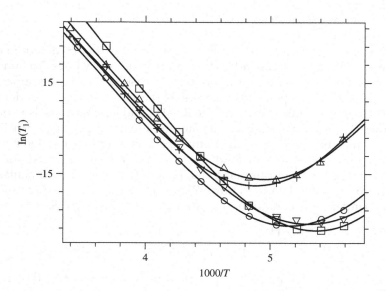

Figure 8.7 Variable-temperature ^1H T_1 data for the hydride resonances in $ReH_2(CO)(NO)(POPr^i_3)_2$; 200 MHz: circles, ReH with $\delta = -1.51$ ppm; downward triangles, ReH with $\delta = -4.78$ ppm; 300 MHz; crosses, ReH with $\delta = -1.51$ ppm; upward triangles, ReH with $\delta = -4.78$ ppm. (Reproduced from D. G. Gusev, D. Nietlispach, A. B. Vymenits, V. I. Bakhmutov, H. Berke. *Inorganic Chemistry*, 1993; **32**: 3270, with permission from Elsevier)

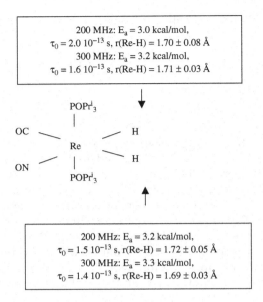

200 MHz: E_a = 3.0 kcal/mol,
τ_0 = 2.0 10^{-13} s, r(Re-H) = 1.70 ± 0.08 Å
300 MHz: E_a = 3.2 kcal/mol,
τ_0 = 1.6 10^{-13} s, r(Re-H) = 1.71 ± 0.03 Å

200 MHz: E_a = 3.2 kcal/mol,
τ_0 = 1.5 10^{-13} s, r(Re-H) = 1.72 ± 0.05 Å
300 MHz: E_a = 3.3 kcal/mol,
τ_0 = 1.4 10^{-13} s, r(Re-H) = 1.69 ± 0.03 Å

Figure 8.8 Re-H bond lengths and parameters of molecular motions (E_a and τ_0) determined for the Re dihydride complex by the variable-temperature ^1H T_1, T_{1sel} relaxation measurements at 200 and 300 MHz in toluene-d_8 [10]

Analysis of Equation (8.10) shows that errors in the $r(M-H)$ determinations by the T_{1sel}/T_1 method depend on the accuracy of the k values. To minimize the errors, the T_{1sel} and T_1 times should be measured in a high-temperature region by several independent experiments at each temperature. Under these conditions, the difference between the Re−H bond lengths in the isomeric Re hydrides (Figure 8.9) is actually meaningful, illustrating a large transligand effect. It must be emphasized again that the very similar activation energies of molecular reorientations (and also the τ_0 values), obtained for both compounds, give strong support to the reliability of these experiments.

According to the data in Figure 8.7, the ^1H T_1 times go through minima (T_{1min}). Under these conditions, Equation (8.10) is rewritten as:

$$r(M-H) = C\,[(1.4k + 4.47)\,T_{1min}/v]^{1/6} \qquad (8.11)$$

$$C = 10^7[\gamma_H^2\,\gamma_M^2\,\hbar^2\,I(I+1)/15\pi]^{1/6}$$

where $r(M-H)$, T_{1min} and v are measured in Å, seconds and MHz, respectively. In the case of Re, Mn, Nb and Ta nuclei, constant C is calculated as 4.20, 4.31, 5.11 and 3.74, respectively. Thus, locations of ^1H T_{1min} times and measurements of ^1H T_{1sel} values in high-temperature zones give $r(M-H)$ via Equation (8.11). Table 8.3 gives examples of such studies carried out in solutions of the niobium hydride systems [11] depicted in Figure 8.10. Owing to the so-called nonclassical interligand Si−H interactions, the SiMe$_2$Cl groups

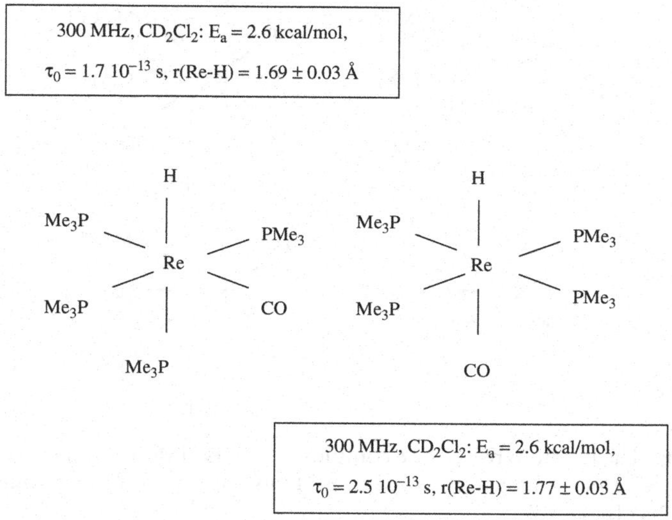
300 MHz, CD2Cl2: Ea = 2.6 kcal/mol,
τ0 = 1.7 10−13 s, r(Re-H) = 1.69 ± 0.03 Å
H
Me3P
PMe3
Re
Me3P
CO
Me3P
H
Me3P
PMe3
Re
PMe3
Me3P
CO
300 MHz, CD2Cl2: Ea = 2.6 kcal/mol,
τ0 = 2.5 10−13 s, r(Re-H) = 1.77 ± 0.03 Å

Figure 8.9 Metal–hydride bond lengths and parameters of molecular motions (E_a and τ_0) measured in two Re monohydrides by the variable-temperature ^{1}H T_1, T_{1sel} relaxation experiments in toluene-d_8 [10]

Table 8.3 Nb–H bond lengths (Å) in complexes I–IV (Figure 8.10) obtained by X-ray (or neutron diffraction, ND) analysis in the solid state, quantum-chemical calculations (DFT), and the $T_{1sel}/T_{1\,min}$ measurements (NMR) in solution (tolene-d_8)

Complex	Hydride	X-ray	DFT	NMR	ND
I	H^A	1.65	1.750	1.79	
	H^X	1.70	1.734	1.80	
		1.76			
II	H^A	1.66	1.793 (1.783)[a]	1.74	
	H^X	1.76	1.745 (1.740)[a]	1.675	
III	H^X	-	1.739 (1.744)[a]	1.714	
IV	H^A	1.74	1.811 (1.791)[a]	1.783	1.816

[a]Different basis set

can affect the Nb–H bond lengths. In accord, the Nb–H bond, neighboring the $SiMe_2Cl$ ligand, is remarkably elongated. This effect is reasonably maximal in complex IV containing two $SiMe_2Cl$ groups. Finally, note that the Nb–H bond lengths in Table 8.3, obtained by NMR relaxation in solution, are longer than those determined by X-ray diffraction in the solid state. In

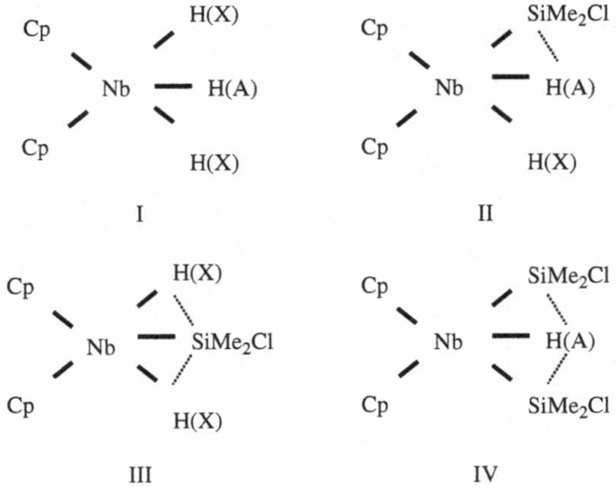

Figure 8.10 The niobium trihydride complexes and its $SiMe_2Cl$ derivatives, illustrating influence of non-classical interligand $Si\cdots H$ interactions on the niobium-hydride bond lengths (Table 8.3)

fact, for objective reasons, the X-ray method underestimates metal–hydride distances while the solution relaxation experiments, DFT calculations and neutron diffraction show very similar data.

Applications of the $^1H\ T_{1sel}/T_{1min}$ method have some restrictions. It is obvious that the inverting 180° pulse will be selective when one of two closely located resonances is excited, but the second is not. Thus, the first restriction can be defined as a minimal chemical shift difference, expressed in Hz, when the acting pulse remains selective. This condition depends on the hardware of NMR spectrometers. However, a crude estimation is 25–30 Hz. The second restriction connects with the nature of the molecules investigated. In fact, the T_{1sel}/T_{1min} approach cannot be applied if a selectively excited resonance is involved in a slow chemical exchange.

8.2 Proton–Proton Distances by Standard $^1H\ T_1$ Measurements

Distances between pairs of protons are accurately determined by standard $^1H\ T_1$ experiments when the corresponding proton–proton dipole–dipole contributions are truly evaluated from the total relaxation rates. If the molecular motion correlation times τ_C are obtained independently, for example, by relaxation experiments on other nuclei, then H–H distances can be calculated via Equation (4.7). The relatively bulky molecular systems often show minimal relaxation times, $^1H\ T_{1min}$. Under these conditions, $r(H-H)$ distances are:

$$r(H-H) = 5.815\ (T_{1min}/\nu)^{1/6} \tag{8.12}$$

where $T_{1 min}$, $r(H-H)$ and ν are measured in seconds, Å and MHz, respectively [6]. On the other hand, these V-shaped plots of $\ln(^1H$ $T_1)$ versus $1/T$ can be fitted to Equation (4.5) to give $r(H-H)$, E_a and τ_0. It is obvious that the fitting procedures are preferable. First, they give a better accuracy in $r(H-H)$ calculations, owing to a larger mass of treated experimental data. Second, good agreements between the theoretical and experimental curves will independently confirm the correctness of an applied dipolar relaxation model. Below we demonstrate different approaches to $r(H-H)$ determinations in solution.

Since dipolar couplings between protons and ^{17}O, ^{13}C and $^{187,189}Os$ nuclei are negligible, proton–proton dipole–dipole interactions make a major contribution to the ^1H relaxation rates in the osmium hydride cluster [3] depicted in Figure 8.11. The ^1H T_1 times, measured by standard inversion recovery experiments, are practically identical for both hydride resonances in toluene-d_8 at 90 MHz (Figure 8.11). This is good evidence for the isotropic character of molecular motions. According to variable-temperature measurements, the ^1H T_1 times vary from 1.67 s (252 K) to 0.73 s (188 K) and go through a minimum of 0.5 s observed at 207 K. The data are well fitted to Equation (4.5) and thus $H-H$ dipole–dipole interactions govern completely the hydride relaxation in the complex, at least at relatively low magnetic fields. Finally the $r(H-H)$, E_a and τ_0 values are very plausible (Figure 8.11) and identical for both hydride resonances.

Hydride resonances of transition metal hydride clusters show significant chemical shift anisotropies ($\Delta\sigma \sim 20$ ppm) in the solid-state ^1H NMR

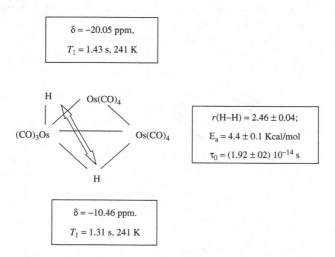

Figure 8.11 The ^1H chemical shifts, the H–H distance and molecular motion parameters determined for the osmium hydride cluster by the variable-temperature ^1H T_1 measurements at 90 MHz in toluene-d_8 [3]

spectra [3]. Since the CSA relaxation rate is proportional to the B_0^2 and $\Delta\sigma^2$ values, the CSA mechanism can become effective at higher magnetic fields. In the case of the Os system in Figure 8.11, the $^1H\,T_{1\,min}$ time, measured at 90 MHz (0.5 s), can be recalculated via Equation (4.9) as 2.22 s at 400 MHz to account for the dipolar nuclear relaxation. However, a $^1H\,T_{1\,min}$ time, obtained experimentally, is shorter (1.93–2.00 s at 227 K). Then, in the presence of the CSA contribution, the total hydride relaxation rate is:

$$1/T_1^{TOT} = 0.3\,(\mu_0/4\pi)^2\,\gamma_H^4\,\hbar^2\,r(H-H)^{-6}$$
$$\times\,[\tau_c/(1+\omega_H^2\tau_c^2)+4\tau_c/(1+4\omega_H^2\tau_c^2)]$$
$$+\,(2/15)\,\gamma_H^2\,B_0^2(\Delta\sigma)^2[\tau_c/(1+\omega_H^2\tau_c^2)] \tag{8.13}$$

Now the $r(H-H)$, E_a, and τ_0 values, previously obtained by relaxation measurements at 90 MHz, can be used for fitting the variable-temperature T_1 data, collected at 400 MHz, to Equation (8.13). In good agreement with the solid-state NMR, the fitting procedure leads to $\Delta\sigma$ values of 22.6 ± 2.0 and 20.0 ± 1.9 ppm for the bridging and terminal hydride ligands, respectively. It is easy to show that the CSA mechanism provides 10% of the total relaxation rate at 400 MHz. This contribution is not large because the H–H distances are overestimated by only 2% if the CSA contribution is not taken into account. However, it is obvious that the errors in the distances will increase at 500, 600 MHz or more and, for this reason, applications of highest magnetic fields in relaxation measurements are often unreasonable. In addition, the highest magnetic fields can lead to the so-called partial alignment of molecular systems. In turn, this effect can rouse anisotropy of molecular motions, perturbing the relaxation behavior. For example, the dihydrogen complex $[Os(H_2)(PPh_3)_2(bpy)(CO)]^+$ shows the $T_{1\,min}(H_2)$ time of 25.2 ms at 750 MHz compared with 22.8 ms, extrapolated from 500 MHz [12].

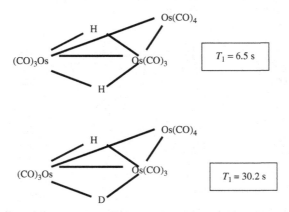

Figure 8.12 $^1H\,T_1$ times of the dihydride osmium cluster and its partially-deuterated derivative measured at 270 MHz and 298 K

Proper proton–proton dipolar contributions can be accurately evaluated by partial deuteration of compounds under investigation. For example, ^1H T_1 measurements at 270 MHz and 298 K in solutions of isotopomeric hydride cluster systems depicted in Figure 8.12 [3], give the proton–proton relaxation rate, calculated as $1/T_1(DD) = (1/6.5 - 1/30.2)(1/0.96)$, where the factor $1/0.96$ compensates the residual dipolar coupling by the ^2H nucleus. In turn, the $1/T_1(DD)$ leads to $r(H–H) = 2.35 \pm 0.03$ Å via Equation (4.7) in excellent agreement with the neutron diffraction studies ($r(H–H) = 2.37$ Å). Note that the τ_C value, necessary for the $r(H–H)$ calculation, has been obtained by ^{13}C and ^{17}O T_1 relaxation experiments on the same compound.

8.3 H–H Distances by T_{1sel}/T_{1bis} Measurements

An alternative method for separation of proton–proton dipole–dipole contributions is based on a combination of the selective (T_{1sel}) and biselective (T_{1bis}) relaxation time measurements [10]. Theoretically, the difference between the $1/T_{1sel}$ and $1/T_{1bis}$ rates gives the *cross-relaxation rate* σ_{ij}:

$$(1/T_{1sel} - 1/T_{1bis})_i = (1/T_{1sel} - 1/T_{1bis})_j = \sigma_{ij}$$

$$\sigma_{ij} = 0.1 \, (\mu_0/4\pi)^2 \, \gamma_H^4 \, \hbar^2 \, r(H_i–H_j)^{-6} \, [6\tau_c/(1 + 4\omega_H^2\tau_c^2) - \tau_c]$$

$$(8.14)$$

where i protons are coupled by j protons. If the molecular motion correlation times τ_c are obtained independently, then σ_{ij} measurements give $H_i–H_j$ distances via Equation (8.14). Figure 8.13 shows the H–H distances found by ^1H T_{1sel}/T_{1bis} experiments on Re complexes in toluene-d_8 solutions [10]. Again,

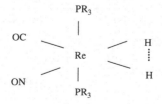

$R = CH_3$, $r(H\cdots H) = 2.25 \pm 0.15$ Å

$R = Cy$, $r(H\cdots H) = 2.34 \pm 0.37$ Å

$R = OPr^i$, $r(H\cdots H) = 2.28 \pm 0.12$ Å

Figure 8.13 The H$\cdots$H distances in the rhenium dihydride complexes determined by the ^1H T_{1sel}/T_{1bis} experiments in toluene-d_8 solutions [10]

the τ_c values, necessary for $r(H-H)$ calculations, have been obtained independently by fitting the variable-temperature ^1H T_1 times to Equation (8.10). As can be seen, the experiments in solution provide good localizations of the hydride ligands, which can be compared with the solid-state data. Actually, to account for the Re—H bond lengths (see above) the H—H distances correspond to H—Re—H angles about 80° as found in the solid state.

To show the magnitudes, measured in the ^1H T_{1sel}/^1H T_{1bis} experiments, and the logic of T_1 interpretations, consider the binuclear Ta/Au hydride complex (see Figure 8.14). Table 8.4 lists the relaxation data obtained for the H^X ligand. It is seen, that the nonselective T_1 time (0.28 s) is remarkably shorter than T_{1sel} (0.38 s). Nevertheless, the T_{1sel}/T_1 ratio is equal to 1.36 compared with 1.50, expected from proton–proton dipolar relaxation. This result is explained by the presence of tantalum–hydride dipole–dipole interactions. The $T_{1sel}(H^X)$ and $T_{1bis}(H^X-Ph)$ times are identical and thus dipole–dipole

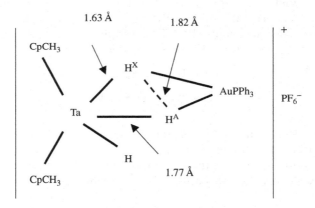

Figure 8.14 Localization of the hydride ligands $(\delta(H^A) = -0.714\,\text{ppm}$, 298 K, $\delta(H^X) = 0.131\,\text{ppm})$ in the binuclear transition metal hydride complex by the T_{1min}, T_{1sel} and T_{1bis} relaxation measurements in acetone-d_6

Table 8.4 T_1 data for the H^X resonance of the Ta/Au hydride complex in acetone-d_6 at 273 K and 200 MHz (CH^α and $CH\beta$ correspond to the α and β protons in the Cp ligands)

Parameter	Value (s)	Parameter	Value (s)
T_1	0.280 (0.385)[a]	$T_{1bis}(H^X-CH^\alpha)$	0.365
T_{1sel}	0.380 (0.528)[a]	$T_{1bis}(H^X-HA)$	(0.442)[a]
$T_{1bis}(H^X-CH_3)$	0.344 (0.488)[a]	T_{1min}[b]	0.0962
$T_{1bis}(H^X-Ph)$	0.383	$T_{1min}(H^X-H^A)$[b]	0.185
$T_{1bis}(H^X-CH\beta)$	0.363		

[a]295 K;
[b]at 210 K

interactions between the H^X ligand and protons in the Ph rings are negligible. The difference between the $T_{1sel}(H^X)$ and $T_{1bis}(H^X-CH^\alpha)$ values is close to the error of the measurements (5%) and hence the corresponding dipolar contribution is also very small. The CH_3 protons provides a more remarkable contribution (compare $T_{1sel}(H^X) = 0.380$ s with $T_{1bis}(H^X-CH_3) = 0.344$ s). However the biggest effect is observed for the H^X/H^A pair where the $T_{1sel}(H^X)$ value is much greater than $T_{1bis}(H^X-H^A)$ (0.528 compared with 0.442 s at 295 K). This result shows that the H^X-H^A dipolar coupling plays a principal role in the H^X relaxation, providing 52% of the total H^X relaxation rate. Finally, the T_1, T_{1sel}, T_{1bis} and T_{1min} data lead to the structure in Figure 8.14.

In contrast to direct T_{1sel}/T_{1bis} measurements in Equation (8.14), the cross-relaxation terms σ_{ij} can be obtained by the one-dimensional NOESY, ROESY (rotating frame Overhauser effect) and TROESY (transfer ROE) experiments [13]. Note that selective excitation of i protons in NOE measurements is realized with the help of shaped pulses of 40–50 ms duration. A spin lock in the TROESY experiments is reached with the locking field where $\gamma B_1/2\pi = 2.8$ kHz. Theoretically, the cross-relaxation rates, measured by the NOE and TROE experiments, are expressed as combinations of the spectral density functions taken at certain frequencies:

$$\sigma_{ij}(\text{NOE}) = (1/4)\,(\mu_0/4\,\pi)^2\,\gamma^4\,\hbar^2\,r(H_i-H_j)^{-6}\,[(6J(2\omega) - J(0)] \tag{8.15}$$

$$\sigma_{ij}(\text{TROE}) = (1/8)\,(\mu_0/4\,\pi)^2\,\gamma^4\,\hbar^2\,r(H_i-H_j)^{-6}\,[(6J(2\omega) + 3J(\omega) - J(0)] \tag{8.16}$$

where $J(0)$, $J(\omega)$ and $J(2\omega)$ are τ_C, $\tau_C/(1 + \omega_H^2\tau_C^2)$ and $\tau_C/(1 + 4\omega_H^2\tau_C^2)$, respectively. Thus, in most cases, determinations of $r(H_i-H_j)$ distances require a knowledge of correlation times τ_C and the character of molecular motions. The latter, as we have shown in Chapter 6, has an influence on choice of the type of the spectral density function.

In the absence of these independent data, unknown proton–proton distances can be obtained by comparison of $\sigma_{ij}(\text{NOE})$, or $\sigma_{ij}(\text{TROE})$, values, measured for a pair of protons, with the cross-relaxation terms, determined for a reference proton pair, where a proton–proton distance is known. According to this so-called isolated spin pair approximation (ISPA), the unknown distances can be calculated via the equation:

$$r(H_i-H_j) = r(\text{REF})\,(\sigma_{\text{REF}}/\sigma_{ij})^{1/6} \tag{8.17}$$

Unknown $r(H_i-H_j)$ distances can be obtained in the framework of the so-called model-free approach. According to this approach, molecular motions are formally described as a combination of slow (and isotropic) global molecular reorientations, characterizing by correlation times τ_M, and additional fast internal motions with correlation times $\tau \ll \tau_M$. Under these conditions, the spectral density function takes the form:

$$J(\omega) = (2/5)\, S^2\, \tau_M/(1 + \omega^2\, \tau_M{}^2) \tag{8.18}$$

where S^2 is the Szabo motional parameter, varying between 0 and 1. It is obvious that at $S^2 = 1$ the internal motion is totally restricted. Again, if the S^2 and τ_M values are known, proton–proton distances are easy calculated from the cross-relaxation rates. Also, in the absence of the S^2 values, this parameter can be inserted into the expression of dipolar relaxation when the measured relaxation rates give unrealistic H–H distances in compounds with known structures.

In many cases, the correlation times of the global and internal motions are comparable. Such motions can be described by the effective correlation times τ_{EFF}. Then:

$$J(\omega) = (2/5)\, \tau_{EFF}/(1 + \omega^2\, \tau_{EFF}{}^2) \tag{8.19}$$

The τ_{EFF} values can be obtained from the $\sigma_{ij}(NOE)$ (or $\sigma_{ij}(TROE)$) terms and $r(H–H)$ distances calculated with the help of the ISPA method. In the absence of the $r(H–H)$ distances, the $\sigma_{ij}(NOE)$ and $\sigma_{ij}(TROE)$ measurements lead to calculations of the τ_{EFF} via:

$$\sigma_{ij}(NOE)/\sigma_{ij}(TROE) = (10 + 42\omega^2\, \tau_{EFF}{}^2 - 32\omega^4\, \tau_{EFF}{}^4 - 32\omega^8\, \tau_{EFF}{}^8)/$$

$$(10 + 63\omega^2\, \tau_{EFF}{}^2 + 96\omega^4\, \tau_{EFF}{}^4 + 16\omega^8\, \tau_{EFF}{}^8) \tag{8.20}$$

All these approaches have been used to determine proton–proton distances in the trisaccharide molecule (Figure 8.15) dissolved in D_2O and $D_2O/DMSO$ mixtures [13]. Figure 8.16 illustrates the TROE, NOE and regular ^1H NMR spectra recorded in a D_2O solution at room temperature. The normalized NOE (and TROE) integral intensities, measured as functions of mixing times under different conditions, are shown in Figure 8.17(a). At short mixing times, the cross-relaxation terms are quantitatively determined by the slopes of linear sections in Figure 8.17(b). The NMR spectra, recorded in a D_2O solution, show positive NOEs and TROEs. In accordance with Figure 4.4,

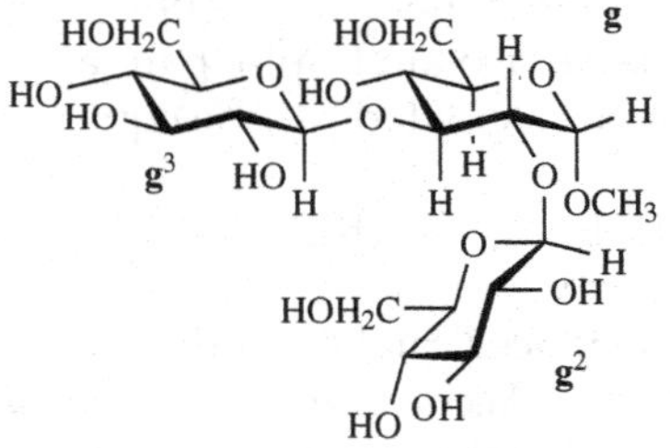

Figure 8.15 The molecular structure of the trisaccharide investigated by the NOE and TROE experiments in D_2O and $D_2O/DMSO$ solutions [13]

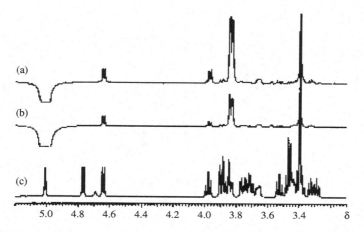

Figure 8.16 TROE (a), NOE (b) and traditional ^{1}H NMR spectra of the trisaccharide (Figure 8.15) in D_2O solution at room temperature (600 MHz). The TROE and NOE spectra are recorded at selective excitation of the H1 proton in the methyl glucoside (g) residue. (Reproduced with permission from T. Rundolf, L. Erikson, G. Widmalm. *Chemistry. A European Journal*, 2001; 7: 1750, © John Wiley & Sons, Inc.)

overall molecular tumbling corresponds to a fast motion regime. In this case, the concept of the effective correlation times can be used to calculate the r(H–H) distances. In contrast, NOEs and TROEs are negative and positive, respectively, in D_2O/DMSO solutions. This situation corresponds to slower molecular tumbling. The calculated proton–proton distances are between 2.14 and 3.27 Å in a good agreement with the X-ray structure of the trisaccharide. Finally, it must be emphasized that errors in the experimental determinations of the cross-relaxation rates are close to 5%. In turn, these errors lead to 1% errors in r(H$_i$–H$_j$) calculations.

8.4 H···H Distances in Intermediates

The dihydrogen-bonded species are good illustrations showing the methodology in studies of intermediates by the NMR relaxation technique. These species are formed when hydride ligands of transition metal hydrides act as proton acceptors in the proton transfer reactions:

$$M-H^{\delta-} + {}^{\delta+}HX \rightleftharpoons MH^{\delta-\cdots\delta+}HX \rightleftharpoons M(H_2)^+ X^- \qquad (8.21)$$

The final products of these reactions are dihydrogen complexes. It is obvious that a key question in characterization of the proton transfer along dihydrogen bond is the H–H bond length. Like normal H-bonds, the H$^{\delta-\cdots\delta+}$H bonding shows a weak or medium strength ($-\Delta H^0$ of $\leqslant$ 7–8 kcal/mol) [14].

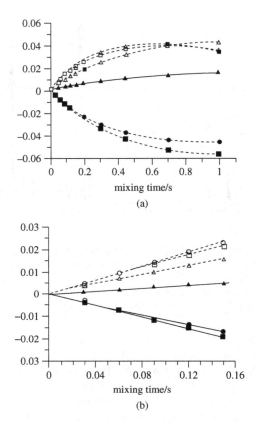

Figure 8.17 Experimental NOE (solid lines) and TROE (dashed lines) curves as a function of mixing time obtained for the H1g/H2g pair and used as the reference interaction in the ISPA distance calculations. (a) TROE curves: open upward triangles, D_2O at 600 MHz; open circles, D_2O/DMSO-D_6 at 500 MHZ; open squares, D_2O/DMSO-D_6 at 600 MHZ; NOE curves; full upward triangle, D_2O at 600 MHz; full squares, D_2O/DMSO-D_6 at 600 MHZ; full circles, D_2O/DMSO-D_6 at 500 MHZ. The corresponding initial linear regions are shown in (b). (Reproduced with permission from T. Rundolf, L. Erikson, G. Widmalm. *Chemistry. A European Journal* 2001; **7**: 1750, © John Wiley & Sons, Inc.)

For these reasons, the formation of $MH^{\delta-\cdots\delta+}HX$ bonds is fast on the NMR time scale, even at the lowest temperatures. In addition, isolation of the $H^{\delta-\cdots\delta+}H$ adducts is very difficult.

^1H NMR can use two target nuclei ($M-^1H$ and ^1H-X) to probe the reversible reactions of Equation (8.21). However the acidic components HX are usually self-associated in solutions and therefore $M-^1H$ resonances are preferable. These resonances are averaged between M^1H and $M^1H^{\delta-\cdots\delta+}HX$ positions, even at the lowest temperatures, while the more energetic formation of dihydrogen complexes can be stopped on the NMR time scale. The latter provides separate observations of the $M(H_2)^+$ resonances. We demonstrate how to determine H−H distances in dihydrogen bonds under these conditions by measuring ^1H T_1 relaxation times.

Figure 8.18, as an example, summarizes the ^1H NMR data collected in a CD_2Cl_2 solution of individual hydride complex (triphos)Ru(CO)H_2 at 200 MHz [15]. The hydride ligands in this complex are magnetically equivalent and show a single ^1H resonance with a relatively long ^1H T_{1min} time. According to the T_1 criterion, this time corresponds closely to the classical dihydride structure. NMR frequencies and a natural abundance of ^{99}Ru nuclei are very low and for these reasons, the ruthenium–hydride dipolar coupling is negligible. Hence, 0.178 s are controlled by hydride–hydride,

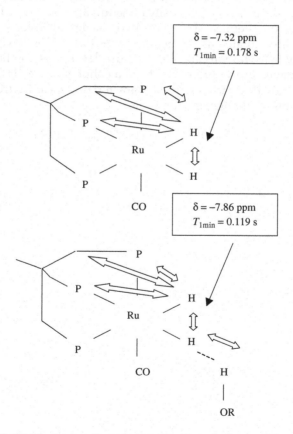

Figure 8.18 ^1H NMR data collected for CD_2Cl_2 solutions of hydride complex (triphos)Ru(CO)H_2 (triphos = MeC(CH$_2$PPh$_2$)$_3$) and its H−H adduct with hexafluoro-2-propanol at 200 MHz and 200 K (arrows show the corresponding dipole–dipole interactions)

hydride–proton and hydride–phosphorus dipole–dipole interactions. It has been established that the formation of dihydrogen bonds is accompanied by high- and low-field shifts of M^1H and ^1HX resonances, respectively [14]. In full accordance with this rule, low-temperature addition of a two-fold excess of hexafluoro-2-propanol (as a proton donor) to a CD_2Cl_2 solution of the Ru hydride causes temperature-dependent high-field shifts of the hydride resonance from -7.32 to -7.86 ppm (Figure 8.19). The plot of the δ versus temperature reaches a plateau at 200 K. It is obvious that equilibrium (Equation 8.21) is completely shifted to the dihydrogen-bonded adduct at this temperature. Note that the hydride chemical shift in individual complex (triphos)Ru(CO)H$_2$ is practically independent of the temperature (Figure 8.19).

The hydride resonance, observed in the presence of the alcohol at 200 K, is averaged between two magnetically nonequivalent and equally populated positions. Then the chemical shift of the hydride ligand, involved in dihydrogen bonding, is calculated as -8.29 ppm via: $\delta(RuH)^{OBS} = 0.5\,\delta(RuH)^{FREE} + 0.5\,\delta(RuH{-}HOR)$. The variable-temperature ^1H T_1 data, collected for the hydride resonance in the presence of the alcohol, show a ^1H T_1 minimum (^1H $T_{1\,min}^{OBS} = 0.119$ s) at 200 K, thus characterizing the dihydrogen-bonded adduct. For a fast exchange one can write:

$$1/T_{1\,min}^{OBS} = 0.5/0.178 + 0.5/T_{1\,min}^{OBS}(RuH{-}H) \qquad (8.22)$$

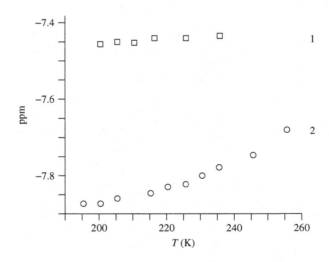

Figure 8.19 Variable-temperature chemical shifts of the hydride resonance in a CD_2Cl_2 of (triphos)Ru(CO)H$_2$ in the absence (squares) and presence (circles) of hexafluoro-2-propanol. (Reproduced with permission from V. I. Bakhmutov *et al. Canadian Journal of Chemistry*. 2001; **79**: 479, © Canadian National Research Council)

Table 8.5 Dihydrogen bond lengths obtained for some hydride systems by the ^1H T_1 relaxation measurements in solutions

System	$r(H-H)$ Å
(triphos)(CO)$_2$Re H$\cdots$HOC(CF$_3$)$_3$	1.83
(triphos)Ru(CO)(H)H...HOCH(CF$_3$)$_2$	1.81
(PMe$_3$)$_2$(NO)(CO)$_2$WH... HOCH(CF$_3$)$_2$	1.77
PP$_3$Os(H)H... HOCH$_2$CF$_3$	1.96

where 0.178 s correspond to the minimal ^1H T_1 time, measured in the individual dihydride, and the $T_{1\,min}^{OBS}$ (RuH$-$H) is a minimal relaxation time of the hydride ligand, involved in dihydrogen bonding. According to Equation (8.22), $T_{1\,min}^{OBS}$ (RuH$-$H) is calculated as 0.0894 s. The obtained value is notably smaller than that in the individual dihydride due to an additional hydride$-$proton dipolar coupling (see Figure 8.18). This additional relaxation rate, $1/T_{1\,min}$(RuH$-$H), governed by the hydride$-$proton dipolar coupling, is expressed by:

$$1/T_{1\,min}(RuH-H) = 1/T_{1\,min}^{OBS}(RuH-H) - 1/0.178 \qquad (8.23)$$

and calculated as 0.183 s. Finally, this value gives a $r(H-H)$ distance of 1.81 Å via Equation (8.12). This distance, being smaller than the sum of the van der Waals radii of H atoms, supports the presence of H$-$H bonding interactions. The same method, applied to other hydride systems, gives similar H$-$H distances (Table 8.5).

8.5 Analyzing the Errors in ^1H T_1 Determinations of Internuclear Distances

As we have shown, exponential nuclear relaxation is treated by standard procedures to give relaxation times with errors of $\leqslant 5\%$. Since rates of dipole$-$dipole relaxation $1/T_1$ are proportional to the inverse *sixth power* of internuclear distances, 5% errors lead to errors in $r(H-H)$ or $r(X-H)$ calculations of less than 1%. These simple estimations are quite valid for molecular systems with isotropic motions. Anisotropic molecular reorientations increase ^1H T_1 times (see Chapter 6) leading to 'effectively' increased internuclear distances calculated on an isotropic assumption. Thus, an incorrect mode of the motion, applied in calculations, results in additional errors. Since, *a priori*, the character of molecular motions is unknown, let us investigate these errors by using, for example, an ellipsoidal model.

Take a symmetric ellipsoidal molecule with a pair of protons, separated by 2 Å, and calculate via Equation (6.5) the ^1H $T_{1\,min}$ times at $\nu_0 = 200$ MHz for ρ values varying from 1 to 2, 5, 10, 50 and ∞. The calculations give ^{1}H $T_{1\,min}$ times of 330, 336, 367, 396, 430 and 1320 ms, respectively. In turn, recalculations of r(H–H) distances in the isotropic approximation (see Equation 8.12) lead to 2.000, 2.005, 2.035, 2.061, 2.089 and 2.518 Å, respectively, compared with the initial value of 2 Å. Compound IrH$_5$(PPr$^i_3$)$_2$ with the bulky trans-located phosphorus ligands can be taken as a typical ellipsoidal molecule. A crude calculation of its moments of inertia I, ignoring the contributions from small hydrogen atoms, gives an $I_\parallel/I_\perp$ ratio of 0.41 [6]. In this connection, the ρ value of 5 can be a reasonable upper limit of the anisotropy. In this case the $T_{1\,min}$ value of 367 ms ($\rho = 5$) corresponds to an effective r(H–H) distance of 2.035 Å. In other words, the H–H distance of 2 Å will be 'observed' as 2.035 Å, producing an error of 2%.

The concept of the correlation time distribution can be also used for estimations of errors in r(H–H) calculations. It has been experimentally established that the symmetric Fuoss–Kirkwood distribution with $\beta = 0.7$ (see Equation 6.8) describes molecular motions of macromolecules, such as polyethylene glycol 200 or polyoxymethylene. Calculations of T_1 relaxation times at 200 MHz via Equation (6.8) for a pair of protons, separated by 2 Å, demonstrate that the ^{1}H $T_{1\,min}$ time increases by 25% on going from $\beta = 1$ (an isotropic model) to $\beta = 0.7$. Even in this situation, the H–H distance of 2 Å will be 'observed' as 2.08 Å, i.e. with an error $\leqslant 4\%$. Thus, in the context of internuclear distance calculations, ^{1}H $T_{1\,min}$ data can be successfully treated by assuming isotropic molecular motions. Finally, it must be emphasized that this important conclusion, allowing us to avoid the problem of how to model the motion, is absolutely valid for r(H–H) calculations only near T_1 minima (see the curves in Figure 6.3). In contrast, for high-temperature regions (i.e. at $1 \gg \omega_H^2\tau_c^2$) the motional anisotropy leads to much greater increases in T_1. It is easy to show that, for example, even at $\rho = 5$ and $1 \gg \omega_H^2\tau_c^2$ the distance of 2 Å will 'observed' as 2.28 Å, leading to an error of 12%.

Bibliography

1. T. Tsang, T. C. Farrar. *Journal of Chemical Physics* 1969; **50**: 3498.

2. M. J. Duer (ed.) *Solid State NMR Spectroscopy. Principles and Applications*. Blackwell: Oxford, 2002.

3. S. Aime, W. Dastru, R. Gobetto, A. Viale. NMR relaxation studies of polynuclear hydride derivatives. In: R. Poli, M. Peruzzini (eds) *Recent Advances in Hydride Chemistry*. Elsevier: Amsterdam, London, New York, Paris, Tokyo, 2001; 351–374.

4. V. I. Bakhmutov, E. V. Vorontsov, G. Boni, C. Moise. *Inorganic Chemistry* 1997; **36**: 4055.

5. V. I. Bakhmutov, E. V. Vorontsov. *Reviews of Inorganic Chemistry* 1998; **18**: 183.

6. P. J. Desrosiers, L. Cai, Z. Lin, R. Richards, J. Halpern. *Journal of the American Chemical Society* 1991; **113**: 4173.

7. T. C. Farrar, G. R. Quinting. *Journal of Physical Chemistry* 1986; **90**: 2834.

8. V. I. Bakhmutov, E. V. Vorontsov, G. I. Nikonov, D. D. Lemenovskii. *Inorganic Chemistry* 1998; **37**: 279.

9. V. I. Bakhmutov, E. V. Vorontsov, M. Peruzzini. *1st Annual HYDROCHEM Meeting*, York, September 2003.

10. D. G. Gusev, D. Nietlispach, A. B. Vymenits, V. I. Bakhmutov, H. Berke. *Inorganic Chemistry*, 1993; **32**: 3270.

11. V. I. Bakhmutov, J. A. K. Howard. *Journal of the Chemical Society Dalton Transactions* 2000; 1631.

12. T. A. Luther, D. M. Heinekey. *Journal of the American Chemical Society* 1997; **119**: 6688.

13. V. I. Bakhmutov, E. V. Vorontsov, E. V. Bakhmutova, G. Boni, C. Moise. *Inorganic Chemistry* 1999; **38**: 1121; T. Rundolf, L. Erikson, G. Widmalm. *Chemistry. A European Journal*, 2001; **7**: 1750.

14. L. M. Epstein, N. V. Belkova, E. S. Shubina. Dihydrogen bonded complexes and proton transfer to hydride ligands by spectral (IR, NMR) studies. In: R. Poli, M. Peruzzini (eds) *Recent Advances in Hydride Chemistry*. Elsevier: Amsterdam, London, New York, Paris, Tokyo, 2001; 391–418.

15. V. I. Bakhmutov, E. V. Bakhmutova, N. V. Belkova, C. Bianchini, L. M. Epstein, D. Masi, M. Peruzzini, E. S. Shubina, E. V. Vorontsov, F. Zanobini. *Canadian Journal of Chemistry* 2001; **79**: 479.

9 Deuterium Quadrupole Coupling Constants from ^{2}H T_1 Relaxation Measurements in Solution

Among numerous quadrupolar nuclei, deuterium is of greatest interest for many reasons. *First*, hydrogen, being one of the most widely distributed chemical elements, is present in compounds belonging to quite different classes. *Second*, the quadrupole moment of ^{2}H nuclei is relatively small and therefore the ^{2}H NMR spectra are well detected. *Third*, relaxation experiments, carried out on two magnetic isotopes, ^{1}H and ^{2}H, provide independent data completely characterizing X–H bonds in terms of bond lengths as well as bonding modes. *Fourth*, direct ^{2}H $T_{1,2}$ relaxation measurements in solutions of the ^{2}H-labeled compounds are standard, simple and convenient. In addition, they do not require special NMR techniques. Even in the absence of a deuterium NMR probe, the ^{2}H relaxation experiments can be performed via a channel of deuterium stabilization.

9.1 How to Determine DQCC Values

Nuclear quadrupole coupling constants and their variation as a function of molecular structure are objects of solid-state investigations by NQR. Application of this method for ^{2}H nuclei is problematic because of their

Practical NMR Relaxation for Chemists   Vladimir I. Bakhmutov
© 2004 John Wiley & Sons, Ltd   ISBNs: 0-470-09445-1 (HB); 0-470-09446-X (PB)

small quadrupole moments. For this reason, the solid-state and solution NMR techniques became a unique tool for determinations and studies of quadrupole parameters at deuterium.

Solid-state NMR [1] provides direct measurements of DQCC values. A typical powder ^2H NMR spectrum of a static sample is shown in Figure 9.1. The shape of the signal exhibits the so-called quadrupolar splitting, marked as Δv. This splitting and a *static* DQC constant, (e^2qQ/h), are interconnected via the relation:

$$\Delta v = 3/4(e^2qQ/h) \tag{9.1}$$

The measurements of the Δv values are particularly convenient for DQCC determinations in the absence of intensive molecular motions occurring on the time scale of quadrupole interactions. Fast molecular motions cause reorientations of eq_{ZZ} vectors, partially (or completely) average quadrupole interactions and strongly affect the shape of ^2H NMR lines. For example, fast

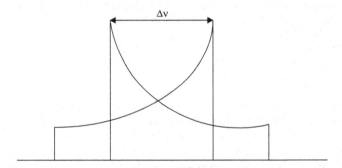

Figure 9.1 Powder static ^2H NMR spectrum of a system, containing a D–X bond with an axially symmetric electric field gradient at deuterium. The spectrum consists of two doublet patterns corresponding to two spin transitions $(0 \rightarrow 1$ and $-1 \rightarrow 0)$

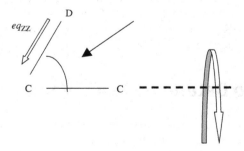

Figure 9.2 A fast molecular rotation around the C–C bond, reorienting the eq_{ZZ} vector and averaging quadrupole interactions at D. The averaging effect depends on the angle formed by the rotational axis and the direction of the principal electric field gradient component, eq_{ZZ}

diffusion around the C−C bond (see Figure 9.2) leads to a decrease of the Δv splitting, observed in solid-state ^{2}H NMR spectra. In this situation, the DQCC value can be calculated via the equation:

$$\Delta v = 3/4(e^2qQ/h)((3\cos^2 \Delta - 1)/2) \qquad (9.2)$$

where Δ is the angle formed by the rotational axis and the direction of the principal electric field gradient component, eq_{ZZ} (Figure 9.2). When the e^2qQ/h frequencies and frequencies of molecular motions are comparable, then the DQCC determination requires the so-called full line shape analysis carried out with the help of specialized NMR programs [2]. The reliability of such calculations depends strongly on the signal-to-noise ratios in the experimental ^{2}H NMR spectra. Note that even in the case of ^{2}H-labeled compounds, the good signal-to-noise ratios in solid-state ^{2}H NMR spectra are often problematic because of the relatively long ^{2}H T_1 times or a low content of deuterium. Nevertheless, solid-state ^{2}H NMR is a unique tool determining the spatial extension of the electric field gradients. In fact, the asymmetry parameter η dictates line shapes of deuterium resonances in the solid state. For example, the spectrum in Figure 9.1 corresponds to an axially symmetric electric field gradient. Asymmetric gradients show shapes that are more complex.

Quadrupolar parameters DQCC and η can be obtained by theoretical calculations. Any molecular orbital calculation results in an electric charge density distribution. Wave functions and nuclear locations give elements of the electric field gradient at deuterium. Finally, a calculated eq_{ZZ} value, expressed in *atomic units*, can be converted into DQC constant For example, DFT (B3LYP/6-31G) computations of deuterium quadrupolar parameters in organic molecules [3] are usually in good agreement with experimental data if DQCC values are calculated with the help of the so-called calibrated deuterium quadrupole moment via the equation:

$$DQCC = eq_{ZZ}(au)\,636.5(kHz\,au^{-1}) \qquad (9.3)$$

The orders of DQCC magnitudes are demonstrated in Table 9.1 where simple organic molecules are taken as examples.

Table 9.1 Experimental and calculated (B3LYP and MP4) deuterium quadrupole coupling constants (kHz) at D in simple organic molecules

Molecule	B3LYP	MP4	Experiment
CF_3D	167.4	152.1	170.8
DCN	204.6	201.9	200.6
DBr	145.7		146.9
DCl	186.5	186.7	188.8
CH_3D	193.1	189.6	191.48
DF	354.7	354.8	354.24

9.2 DQCC Values from ^{2}H T_1 Measurements in Solution (Fast Motion Regime)

Quadrupole interactions completely dominate relaxation of nuclei with spins $>1/2$ when charge distributions around these nuclei are asymmetric. In the case of deuterium, this statement can be verified by experiments on the iso-topomeric hydride/dihydrogen systems depicted in Figure 9.3 [4]. Hydride regions of ^{2}H and ^{1}H NMR spectra of these transition metal hydride complexes show two well-resolved resonances and thus T_1 times can be accurately measured for the terminal and dihydrogen ligands. Owing to a very short D—H distance (0.92 Å) in the (D—H) ligand, deuteron–proton dipole–dipole interactions could give a remarkable contribution to the (D—H) relaxation rate in the partially deuterated complex $PP_3Ru(DH)D$ [4]. However, in spite of the short internuclear distance, the ^{2}H T_1 times, measured for the (D_2) and (DH) ligands, are identical within the limits of error (Table 9.2). That is why the ^{2}H T_1 relaxation is applied as an excellent tool for studies of quadrupole parameters and molecular motions in solutions, liquid crystals and the solid state.

In a high-temperature zone (i.e. at $1 \gg \omega_Q^2\tau_c^2$), the rate of deuterium relaxation, $1/T_1$, is:

$$1/T_1 = 1.5\pi^2(\mathrm{DQCC})^2(1 + \eta^2/3)\tau_c \qquad (9.1)$$

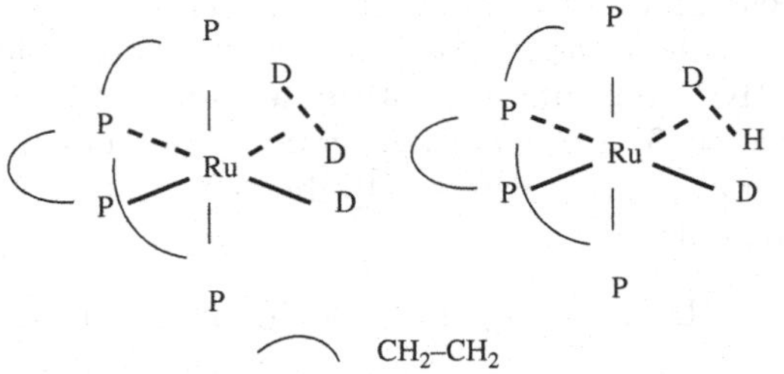

Figure 9.3   The rhenium hydride/dihydrogen complex and its partially-deuterated derivative, illustrating complete domination of quadrupole interactions in the deuterium relaxation

Table 9.2   Variable-temperature deuterium T_1 relaxation times (s) measured in a CH_2Cl_2 solution of isotopomers PP_3RuD_3 and $PP_3Ru(DH)D$ (Figure 9.3) at 61.402 MHz

Complex	$Ru(D_2)$ or $Ru(D-H)$	RuD	$T(K)$
PP_3RuD_3	0.0586	0.0165	180
$PP_3Ru(DH)D$	0.0545	0.0145	180
PP_3RuD_3	0.0780	0.0158	200
$PP_3Ru(DH)D$	0.0815	0.0168	200

To determine DQCC values from $1/T_1$, the molecular motion correlation times τ_c should be found independently, for example, by relaxation experiments on other nuclei (^1H, ^{13}C etc.) located in the same molecule. It is obvious that the asymmetry parameter η cannot be determined via the relaxation data. By definition, this parameter varies between zero and one. Therefore even in the absence of a proper knowledge of the η, an uncertainty in DQCC values, calculated from ^2H T_1 times, is $\leqslant 10\%$. In the case of simple chemical bonds D−X the ambiguity in the η is significantly less. Actually in most such bonds (for example, CD_3X (X=F, Cl, Br, I), CD_3CN and D_2CO [4, 5]) the electric field gradient tensors at D are practically axially symmetric ($\eta < 0.2$). In addition, the principal components eqz_{ZZ} usually lie along these bonds.

Figure 9.4 shows the data collected for the ^2H-labeled Os cluster in toluene-H_8 at 223 K. The ^2H T_1 times of the terminal and bridging hydride ligands are different [6]. Anisotropic molecular tumbling (see Chapter 6) could cause this effect. However, the T_1 times, measured for both ligands at the ^1H frequency, are identical (see Figure 8.11) and thus motions of the cluster seem to be isotropic. Parameters of the motions, E_a and τ_0, found by the variable-temperature ^1H T_1 measurements (Figure 8.11), lead to calculation of a τ_C value at 223 K. The latter, by assuming $\eta = 0$ in Equation (9.1), results in DQCC values of 86.4 ± 1.5 and 60.1 ± 2.0 kHz for the terminal and bridging D-ligands, respectively. According to the DFT calculations of the Ru, W and Os hydrides [7] (Table 9.3) and the solid-state ^2H NMR spectra of complexes $MnD(CO)_5$, Cp_2MoD_2, Cp_2WD_2 and Cp_2ZrD_2, the electric field gradient tensors at the terminal D ligands are actually axially symmetric. In contrast, the tensors at bridging D ligands can deviate significantly from axial symmetry. For example, $\eta = 0.31$ in binuclear systems $[R_4N][DMr_2(CO)_{10}]$ (M=Cr, W) [5]. However even at $\eta = 0.31$ the ^2H T_1 time of the Os−D−Os

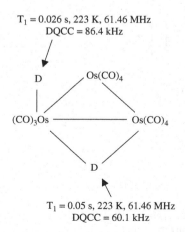

$T_1 = 0.026$ s, 223 K, 61.46 MHz
DQCC = 86.4 kHz

$T_1 = 0.05$ s, 223 K, 61.46 MHz
DQCC = 60.1 kHz

Figure 9.4 ^2H T_1 times and deuterium quadrupole coupling constants measured for the hydride ligands in the ^2H-labeled Os cluster in toluene-H_8 at 223 K

ligand (0.05 s) gives DQCC = 63.3 kHz (instead of 60.1 kHz, obtained in the $\eta = 0$ approximation).

Figure 9.5 illustrates another approach to DQCC determinations, based on T_1 times, measured for reference ^{2}H nuclei. The Ru tetrahydride represents a hydride/dihydrogen system where the terminal and (H$_2$) ligands undergo a

Table 9.3 Major components of the electric field gradient tensors eq_{ZZ} and the asymmetry parameters η calculated by the DFT (B3LYP) method

Compound	eq_{ZZ} (au)	η
D$_2$	−0.397	0.01
WD(NO)(CO)$_2$(PH$_3$)$_2$	−0.1044	0.078
OsD(D$_2$)(CO)Cl(PH$_3$)$_2$	−0.1830	0.085
PP$_3$Ru(D$_2$)D$^+$	−0.1314	0.056
PP$_3$Os(D$_2$)D$^+$	−0.1490	0.053

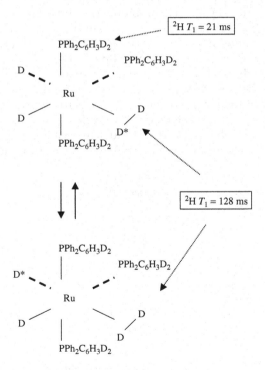

Figure 9.5 T_1 times, measured for Ru—D and ortho-D resonances of the aromatic rings in a toluene-H$_8$ solution of the Ru hydride/dihydrogen complex at 290 K and 30.7 MHz. Under these conditions, the complex undergoes a fast hydride/dihydrogen exchange, leading to the observation of an averaged hydride resonance

fast exchange on the NMR time scale. Under D_2 atmosphere, *ortho*-protons of the aromatic rings and the Ru−H hydrogens change for deuterium [8]. Both resonances are well observed in the room-temperature ^2H NMR spectrum and show different ^2H T_1 times. At $\eta = 0$, the ^2H T_1 times depend on DQCC values at D in the C−D and Ru−D bonds. Since the quadrupole coupling constant for aromatic deuterons is known from solid state ^2H NMR (182 KHz), the DQCC^{D-Ru} value is calculated as 68.7 ± 3.0 KHz.

Farrar *et al.* [9, 10] have suggested an alternative method for measurements of DQCC values in liquids. This indirect method can be applied for molecular systems containing a deuterium capable of hydrogen bonding (for example, the OD or ND bonds). For systems with OD groups, it has been established that the H-bond formation affects the deuterium quadrupole coupling constant that is related to the hydrogen bond distance as:

$$DQCC(OD) = 310 - 600/r^3{}_{OD-O} \tag{9.2}$$

where 310 is the DQCC(OD) value in the gas phase. Farrar calculated (*ab initio*) chemical shifts and DQCC values in H-bonded clusters of methanol and showed that these magnitudes correlate:

$$DQCC(OD)(kHz) = 284 - 15.3\,\delta(OH)(ppm) \tag{9.3}$$

Thus, the DQCC value can be found via the experimentally measured chemical shifts, $\delta(OH)$, for example, in CCl_4 solutions with a variable concentration of methanol. Then, the ^2H $T_1(OD)$ times, measured in these solutions, and the calculated DQCC(OD) values give the correlation time τ_C as a function of concentration C (see Table 9.4). If viscosities of the solutions are measured independently, the Stokes−Einstein−Debye model, or other models [9] allow one to estimate the sizes of H-bonded clusters (see parameter a in Equation 1.13).

On the other hand, Farrar's method, applied for molecular systems with two or more ^2H-labels, allows one to establish the character of molecular tumbling in solution. For example, DQCC values, determined by this method

Table 9.4 Viscosities η, relaxation times T_1, chemical shifts $\delta(OH)$, deuterium quadrupole coupling constants DQCC, and the correlation times τ_C, for various mole fractions C, of methanol in CCl_4

C	η (cp)	T_1 (s)	$\delta(OH)$ (ppm)	DQCC (kHz)	τ_C (ps)
0.0026	0.888	1.720	0.35	278	0.51
0.012	0.886	0.700	0.83	271	1.31
0.029	0.882	0.308	1.76	257	3.33
0.209	0.860	0.172	4.38	217	8.38
0.504	0.828	0.170	4.72	211	8.89
1.000	0.537	0.292	4.81	210	5.25

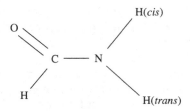

Figure 9.6   The structure of formamide, undergoing anisotropic molecular motions

in liquid formamide (Figure 9.6) led to correlation times τ_C, which are different for reorientations of the *cis*-N−D, *trans*-N−D and C−D vectors: 6.3, 9.2 and 5.2 ps, respectively (295 K). It is obvious that molecular tumbling in neat liquid formamide is anisotropic.

9.3   DQCC Values via ^{2}H $T_{1\,min}$ Measurements in Solution

Theoretically, the plots of $\ln(T_1(Q))$ versus $1/T$ are V-shaped and go through minima at $\tau_c = 0.62/\omega_Q$. Under these conditions, Equation (4.23), written for deuterium relaxation, converts to:

$$DQCC = 1.2201\ (1 + \eta^2/3)^{-1/2}(\nu_D/T_{1\,min})^{1/2} \qquad (9.4)$$

where ν_D, DQCC and $T_{1\,min}$ are measured in MHz, kHz and seconds, respectively. Thus, at $\eta = 0$, ^{2}H $T_{1\,min}$ times give DQCC values directly.

   In solution, the minimal T_1 times can be located by the variable-temperature relaxation experiments for molecular systems with relatively slow motions on the scale of NMR frequencies. This situation is realized in solutions of bulky compounds or viscous media, studied at ^{1}H NMR frequencies. For example, dilute solutions of transition metal hydride complexes show ^{1}H $T_{1\,min}$ times at temperatures between $-60°$ and $-90°$C and at ^{1}H NMR frequencies of 300–500 MHz. It is obvious that, owing to the smaller gyromagnetic ratio of deuterium relative to hydrogen, ^{2}H $T_{1\,min}$ times can be reached for the same complexes at significantly lower temperatures. It is easy to show that relaxation times of molecules, placed in a magnetic field of 9.39 T (i.e. $\nu_H = 400$ MHz and $\nu_D = 61.4$ MHz), will be minimal at $\tau_c = 2.5 \times 10^{-10}$ and 16.0×10^{-10} s for protons and deuterons, respectively. If a ^{1}H $T_{1\,min}$ time for a molecule is observed, for example, at $-70°$C and the activation energy of molecular reorientations is 2.5 kcal/mol, then a ^{2}H $T_{1\,min}$ time for the same molecule can be observed only at $-117°$C. Smaller activation energies will require lower temperatures which are, however, restricted by the freezing points of regular organic solvents. Applications of organic solvents in mixtures, such as, for example, $CH_2Cl_2 : CFCl_3$ in a 1:1 ratio, can extend the diapason of the temperature experiments. Another approach is based on investigations of *concentrated* solutions. Owing to the higher viscosity of such systems, the effective τ_C and E_a values increase. For this reason, $T_{1\,min}$ locations are shifted towards higher temperatures. For example, a *concentrated* CD_2Cl_2 solution of the hydride complex $[Ru(H)(H_2)(dppe)_2]$ $[BPh_4]$ shows a ^1H $T_{1\,min}$ time at 274 K (Table 9.5). The ^1H T_1 minimum in a *dilute* solution of the same system is located at 40° lower [11]. Similar effects are observed in toluene solutions of the dihydride $(PCy_3)_2ReH_2(NO)(CO)$ [12]. In addition, it should be emphasized that the $T_{1\,min}$ times, measured in dilute and concentrated solutions (or in concentrated solutions of different solvents), are practically identical. This experimental fact is a very important argument for quantitative interpretations of $T_{1\,min}$ times measured in concentrated solutions.

The ^2H $T_{1\,min}$ times and DQC constants, determined in solutions of transition metal hydride complexes, are collected in Table 9.6. As can be seen, DQCC values (varying between 55 and 87 kHz) depend on the nature of metals and ligand environments. The chemical meaning of DQCC variations is not a subject of this book. Note however that the DQC constant can be used as a measure of the ionic character of D–X bonds [4, 12]. According to such an interpretation, for example, the W–D bond in complex *trans*-$WD(CMes)(dmpe)_2$ shows a surprisingly high ionicity ($i = 0.84–0.85$) close to that in the LiD molecule. The latter is independently supported by an unusual reactivity of the W–D bond.

Table 9.5 Conditions for locations of minima in variable-temperature $T_{1\,min}$ curves obtained at proton and deuterium frequencies in dilute (D) and concentrated (C) solutions of the hydride complexes

Complex	$T_{1\,min}/\nu$ (s/MHz)	T(K)	Solution
$(PCy_3)_2ReH_2(NO)(CO)$	^1H: 0.0960/200[a]	207	D, toluene-d_8
	^1H: 0.0936/200[a]	206	
	^1H: 0.144/300	243	C[b], toluene-d_8
	^2H: 0.0142/46.04[a]	203	C, toluene-h_8
	^2H: 0.0154/46.04[a]	203	
	^2H: 0.0139/46.04[a]	183	C, CH_2Cl_2
	^2H: 0.0147/46.04[a]	183	C, CH_2Cl_2
$[Ru(H)(H_2)(dppe)_2]^+$	^1H: 0.020/400	203	D, CD_2Cl_2
	^1H: 0.020/400	273	C, CD_2Cl_2
	^2H: 0.181/61.45	193	C, CH_2Cl_2

[a]Measured for two different resonances
[b]100–150 mg/ 1 ml

Table 9.6 DQCC values for terminal hydride ligands in transition metal hydride complexes. Data obtained via Equation (9.4) from ^2H T_{1min} times measured in solution

Compound	^2H T_{1min} (ms)/ DQCC (kHz)	ν_D (MHz)	T(K)	Solvent
[Ru(D)(D$_2$)(dppe)$_2$][BPh$_4$]	15/79	61.45	193	CH$_2$Cl$_2$
[Os(D)(D$_2$)(dppe)$_2$][PF$_6$]	14/81	61.45	193	CH$_2$Cl$_2$
[(triphos)RhD$_3$]	16.5/83.2	76.75	190	CH$_2$Cl$_2$
[(triphos)IrD$_3$]	10.1/95.0	61.45	190	CH$_2$Cl$_2$
[Os(D)(D$_2$)(CO)(Cl)(PPr^i$_3$)$_2$]	9.0/87.3	46.04	195	Toluene-H$_8$
[PP$_3$RhD$_2$][CF$_3$COO]	21.7/73.9	76.75	220	THF
[PP$_3$RhD$_2$]	20.2/75.2	76.75	190	THF
[PP$_3$CoD$_2$]	19.5/76.5	76.75	210	THF
(PMe$_3$)$_2$ReD$_2$(NO)(CO)	14.0/70.0	46.04	173	Toluene-H$_8$
	16.1/65.3	46.04	173	Toluene-H$_8$
(PMe$_3$)$_2$WD(NO)(CO)$_2$	22.7/55.0	46.04	178	Toluene-H$_8$
(PPh$_3$)$_2$WD(NO)(CO)$_2$	22.5/55.2	46.04	203	Toluene-H$_8$
(PEt$_3$)$_2$MnD(NO)$_2$	21.5/56.4	46.04	186	Toluene-H$_8$
(PEt$_3$)$_2$MnD(CO)$_3$	15.4/66.7	46.04	183	Toluene-H$_8$
cis-(PMe$_3$)$_4$ReD(CO)	12.4/74.4	46.04	163	CH$_2$Cl$_2$
trans-(PMe$_3$)$_4$ReD(CO)	15.3/66.8	46.04	163	CH$_2$Cl$_2$

9.4 Errors in DQCC Determinations

Deuterium relaxation in solutions of compounds with relatively low molecular weights (non-polymer molecular systems) is monoexponential and therefore ^2H T_{1min} times are determined with errors $\leqslant 5\%$. In turn, the rate of nuclear relaxation is proportional to the *second power* of the quadrupole coupling constant. If molecular tumbling is isotropic, then calculations via Equations (9.1) and (9.4) give DQCC values with errors $\leqslant 2.2\%$. Larger errors can appear when an isotropic approximation is applied for systems with anisotropic motions. By analogy with the ^1H dipole–dipole relaxation in symmetric ellipsoidal molecules (see Chapters 6 and 8), these errors are particularly significant in high- and low-temperature zones where they depend strongly on the motional anisotropy, ρ (see Figure 6.3). In contrast, the errors in T_{1min} zones are smaller and they are less sensitive to the motional anisotropy. To make the considerations more quantitative, take $\rho = 5$ as a reasonable *upper* limit of the motional anisotropy for regular compounds. Then it is easy to show that on going from $\rho = 1$ (isotropic motions) to $\rho = 5$ the ^2H T_{1min} time increases by 10%, leading an effective DQCC value reduced by 5–6%. Thus, if the character of molecular tumbling is unknown, a better accuracy cannot be reached, even in DQCC determinations based on ^2H T_{1min} times.

Figure 9.7 shows the variable-temperature ^2H T_1 data, collected in a concentrated toluene-H$_8$ solution of the complex ReD$_2$(CO)(NO)(PCy$_3$)$_2$. As it

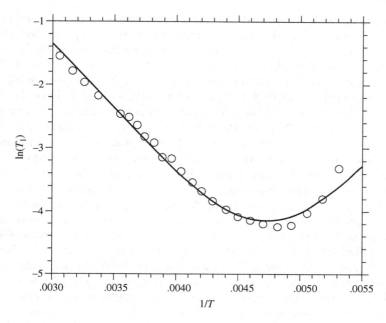

Figure 9.7 Variable-temperature ^2H T_1 data, ln (T_1) versus $1/T$, collected for the D ligands of complex $ReD_2(CO)(NO)(PCy_3)_2$ in a concentrated toluene-H_8 solution at 46.06 MHz. (Reproduced with permission from D. Nietlispach, V. I. Bakhmutov, H. Berke. *Journal of the American Chemical Society* 1993; **115**: 9191. © 1993 American Chemical Society)

is seen, the ^2H T_1 times deviate from the symmetrical V-shaped curve (the solid line), obtained for isotropic molecular tumbling. The deviations are particularly remarkable in the $T_{1\,min}$ range and the low-temperature section. Such effects are typical of solids, glasses and viscous liquids and are usually interpreted in terms of correlation time distributions [13]. However, it is interesting that the ^1H relaxation in dilute solutions of $ReH_2(CO)(NO)(PR_3)_2$ corresponds well to the isotropic behavior (see Figure 8.7). In this connection, we do not consider a correlation time distribution as a real physical origin of the deviations, observed in concentrated solutions (Figure 9.7). We use this concept as a convenient mathematical model for estimations of errors in DQCC determinations. If an X–D bond, having a DQCC value of 100 KHz and an axially symmetric electric field gradient at D ($\eta = 0$), is placed into a magnetic field of 7.05 T ($\nu_D = 46.06$ MHz), then in the presence of the symmetric Fuoss–Kirkwood distribution the ^2H T_1 relaxation rate is [13]:

$$1/T_1 = \{2.958\,(DQCC)^2\beta/(2\pi\nu_D)\}\{(2\pi\nu_D\tau_C)^\beta/(1 + (2\pi\nu_D\tau_C)^{2\beta})$$
$$+ 2(4\pi\nu_D\tau_C)^\beta/(1 + (4\pi\nu_D\tau_C)^{2\beta}\} \qquad (9.5)$$

where β is the width of the distribution, T_1 is measured in seconds and DQCC and ν_D is expressed in Hz. At $\beta = 1$, Equation (9.5) corresponds to isotropic motions and gives $^2H\ T_{1\,min} = 6.86$ ms. Decreasing the β parameter from 1 to 0.9, 0.8 and 0.7 results in an increase of the $^2H\ T_{1\,min}$ time to 7.54, 8.42 and 9.55 ms, respectively. These times, treated in an isotropic approximation (i.e. via Equation (9.4)), give DQCC values of 95.4, 90.1 and 84.7 kHz versus initial 100.0 kHz. In other words, the deuterium quadrupole coupling constants are underestimated by 5, 10 and 18% with increasing width of the distribution.

The best way to determinations of DQC constants in the presence of correlation time distributions is fitting the experimental T_1 times to the corresponding models of the motion. It is obvious that the method requires a large mass of experimental data, particularly in low-temperature regions. However, the latter is problematic for liquids and solutions where low-temperature experiments are restricted by freezing points. In this situation, reliability of the fitting procedures is questionable, particularly in the absence of a detailed knowledge of molecular dynamics in concentrated solutions. For estimations of possible β values, affecting final results, consider the data available in the literature. Nuclear relaxation in 1,2,3,4-tetrahydro-5,6-dimethyl-1,4-methanonaphthalene, actually classified as an *organic glass-forming liquid*, deviates from the model of isotropic motions. Treatments of the relaxation data in the framework of a correlation time distribution lead to the β values lying between 0.9 and 0.8 [14]. Viscous glycerol also shows a quite narrow correlation time distribution with $\beta = 0.97$ [15]. According to the $^2H\ T_1$ data (Figure 9.7), treated in an isotropic approximation, molecules of $(PCy_3)_2ReD_2(NO)(CO)$ are reoriented in a concentrated toluene solution with an activation energy of 4.0 kcal/mol [12]. Multinuclear relaxation, providing determination of the $\tau_C(T_{1\,min})$ values at ^1H, ^2H, ^{13}C and ^{31}P frequencies (Figure 9.8), result in a slightly larger E_a value (4.6 kcal/mol). Thus if the symmetric Fuoss–Kirkwood correlation time distribution is actually present, then its β value is calculated as: $4.0/4.6 = 0.87$. Fitting the $^2H\ T_1$ times, measured in a toluene solution of the Os cluster (see Figure 9.4) [16], to Equation (9.5) also gives a narrow correlation time distribution: $\beta = 0.94$. All these data allow one to assume that, in spite of the absence of detailed knowledge of molecular dynamics in concentrated solutions, DQCC values, calculated via an isotropic approximation, can be underestimated by only 5–6%. This conclusion is supported by good agreements between the solution and solid-state data in Table 9.7. For example, a concentrated $CHCl_3$ solution of ruthenium hydride $[RuD(\eta^6\text{-toluene})(Binap)]\ [CF_3SO_3]$ shows a $^2H\ T_{1\,min}$ time of 16 ms at 76.8 MHz. The corresponding DQCC value is calculated via Equation (9.4) as 85 kHz. The ^2H MAS NMR spectra of the solid hydride, recorded at 76.8 MHz and at different spinning frequencies, give 89 kHz. As can be seen, even in this case the difference is less than 5%.

Finally, it should be added that activation energies of molecular motions in viscous liquids can become temperature dependent. In other words, the

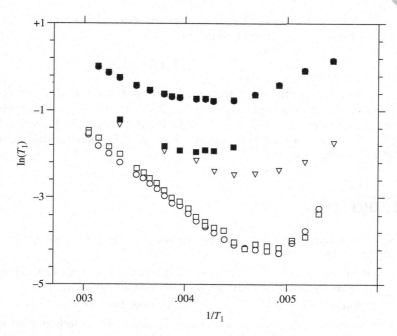

Figure 9.8 Variable-temperature T_1 data (T_1 in seconds) for a toluene solution of complex $ReD_2(CO)(NO)(PCy_3)_2$: solid circles, ^{31}P, 121.4 MHz; solid squares, ^1H-Re, 300 MHz; triangles, $^{13}CH_2$, 75.4 MHz; open circles, 2H-Re, 46.04 MHz. (Reproduced with permission from D. Nietlispach, V. I. Bakhmutov, H. Berke, *Journal of the American Chemical Society* 1993; **115**: 9191. © 1993 American Chemical Society)

Table 9.7 DQCC values (kHz), determined by the ^2H T_1 technique in solution and the solid-state ^2H NMR spectra[a]

Compound	Solution	Solid State
trans-WD(CMes)(dmpe)$_2$	34.1	34.8
(PMe$_3$)$_2$WD(NO)(CO)$_2$	55.0	
Cp$_2$WD$_2$		54
(PEt$_3$)$_2$MnD(NO)$_2$	56.4	
(PEt$_3$)$_2$MnD(CO)$_3$	66.7	
MnD(CO)$_5$		68.1
[Ru(**D**)(D$_2$)(dppe)$_2$][BPh$_4$]	79.0	75
CD$_2$ groups	164[b]	167
[RuD(η^6-toluene)(Binap)][CF$_3$SO$_3$]	85	89

[a]DQCC values calculated from the quadrupolar splitting via Equation (9.1)
[b]Found for CD$_2$ groups in the (dppm) ligand of complex [Cp*Ru(D)(Me)(dppm)]

E_a can increase on cooling. Theoretically, the effect is expressed via the Vogel–Fulcher–Tamman (VFT) equation:

$$\tau_C = \tau_{VFT}\exp\{E_{VFT}/[R(T - T_0)]\} \qquad (9.6)$$

where T_0 is close to the glass transition temperature. Good examples of this behavior are viscous glycerol and 1,2,3,4-tetrahydro-5,6-dimethyl-1,4-methanonaphthalene [14, 15]. It is obvious that application of Equation (9.6) instead the usual exponential function can be used to increase the accuracy of DQCC calculations.

Bibliography

1. M. J. Duer (ed.) *Solid State NMR spectroscopy. Principles and Applications.* Blackwell: Oxford, 2002.

2. F. Wehrmann, J. Albrecht, E. Gedat, G. J. Kubas, J. Eckert, H. H. Limbach, G. Buntkowsky. *Journal of Physical Chemistry A* 2002; **106**: 1977.

3. W. C. Bailey. *Journal of Molecular Spectroscopy* 1998; **190**: 318.

4. V. I. Bakhmutov. Deuterium spin lattice relaxation and deuterium quadrupole coupling constants. A novel strategy for characterization of transition metal hydrides and dihydrogen complexes in solution In: M. Gielen, R. Willem, B. Wrackmeyer (eds.) *Unusual Structures and Physical Properties in Organometallic Chemistry.* Wiley: Chichester, 2002; 145–165.

5. L. G. Butler, E. A. Keiter. *Journal of Coordination Chemistry* 1994; **32**: 121.

6. S. Aime, W. Dastru, R. Gobetto, A. Viale. NMR relaxation studies of polynuclear hydride derivatives. In: R. Poli, M. Peruzzini (eds.), *Recent Advances in Hydride Chemistry.* Elsevier: Amsterdam, London, New York, Paris, Tokyo, 2001; 351–374.

7. V. I. Bakhmutov, C. Bianchini, F. Maseras, A. Lledos, M. Peruzzini, E. V. Vorontsov. *European Journal of Chemistry* 1999; **5**: 3318.

8. D. G. Gusev, A. B. Vymenits, V. I. Bakhmutov. *Inorganica Chimica Acta* 1991; **179**: 195.

9. M. A. Wendt, T. C. Farrar. *Molecular Physics* 1998; **95**: 1077.

10. M. J. Hansen, M. A. Wendt, T. C. Farrar. *Journal of Physical Chemistry A* 2000; **104**: 5328.

11. G. A. Facey, T. P. Fong, D. G. Gusev, P. M. Macdonalds, R. H. Morris, M. Schlaf, W. Xu. *Canadian Journal of Chemistry* 1999; **77**: 1899.

12. D. Nietlispach, V. I. Bakhmutov, H. Berke. *Journal of the American Chemical Society* 1993; **115**: 9191.

13. P. A. Beckmann. *Physics Reports* 1988; **171**: 85.

14. L. Sturz, A. Dolle. *Journal of Physical Chemistry A* 2001; **105**: 5055.

15. A. Friedrich, A. Dolle, M. D. Zeodler. *Magnetic Resonance in Chemistry* 2003; **41**: 813.

16. S. Aime, W. Dastru, R. Gobetto, A. Viale. *Inorganic Chemistry* 2000; **39**: 2422.

17. T. J. Geldbach, H. Ruegger, P. S. Pregosin. *Magnetic Resonance in Chemistry* 2003; **41**: 703.

10 Spin – Lattice ^{1}H and ^{2}H Relaxation in Mobile Groups

In the previous chapters, we have shown that nuclear relaxation in solutions of molecular systems undergoing isotropic motions (or systems with an insignificant motional anisotropy) is a reliable tool for determinations of internuclear distances and quadrupole coupling constants. Vice versa, known internuclear distances or deuterium quadrupole coupling constants provide accurate predictions of ^{1}H or ^{2}H T_1 times. However, this situation changes if target nuclei are more mobile than whole molecules and undergo additional motions, which are fast on the time scale of molecular tumbling. Phenomenologically such motions partially average dipolar (or quadrupolar) coupling without a reorientation of the whole molecule. As a result, the observed T_1 time increases. Table 10.1 shows the minimal ^{1}H T_1 times ($T_{1 min}$(OBS)), measured in solutions of two polyhydride molecules at 500 MHz. As can be seen, the $T_{1 min}$ times (^{1}H $T_{1 min}$(calc)), calculated on the basis of the H–H and H–P distances in the neutron diffraction structures [1], are significantly shorter. Note that the calculations have been performed for 'immobile' hydride ligands and the term 'immobile' corresponds to a situation when the hydrogen atoms participate only in molecular tumbling. It is easy to show that simple anisotropic motions of the hydride molecules cannot explain the large disagreements in Table 10.1. In the case of the iridium pentahydride, an alternative reason for a four-fold T_1 difference could

Practical NMR Relaxation for Chemists   Vladimir I. Bakhmutov
© 2004 John Wiley & Sons, Ltd   ISBNs: 0-470-09445-1 (HB); 0-470-09446-X (PB)

Table 10.1   Calculated and experimental minimal ^1H T_1 relaxation times (s, 500 MHz) obtained for the hydride ligands in two classical hydrides, $IrH_5(PPr^i_3)_2$ and $[ReH_8(PPh_3)]^-$

Hydride	$T_{1\,min}(calc)$	$T_{1\,min}(obs)$	$T_{1\,min}(obs)/T_{1\,min}(calc)$
$IrH_5(PPr^i_3)_2$	0.161	0.599	3.72
$[ReH_8(PPh_3)]^-$	0.121	0.490	4.04

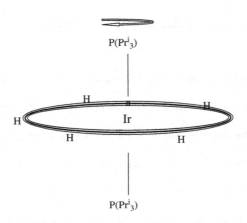

Figure 10.1   Motions of the hydride ligands as a fast 'autonomous rotation' of the polyhydride ring, causing a four-fold elongation of ^1H T_1 times in the iridium hydride

be a fast (on the scale of molecular tumbling) 'autonomous rotation' of the polyhydride ring [2, 3], shown in Figure 10.1. The vectors, connecting pairs of hydrogen atoms, are perpendicular to the rotational axis and then the Woessner equation predicts a four-fold ^1H T_1 increase.

Both examples illustrate the high sensitivity of nuclear relaxation to fast internal motions. The purpose of this chapter is to show how the fast internal motions affect internuclear distances (or deuterium quadrupolar parameters), calculated on the basis of a simple isotropic model, and how to modify the spectral density functions to account for such motions.

One of the simplest molecular models illustrating a high intramolecular mobility is a pair of hydrogen atoms, binding to a metal center in the W dihydrogen complex $W(H_2)(CO)_3(PPr^i_3)_2$ (Figure 10.2). According to the neutron diffraction data [4], the hydride atoms in the complex are separated by 0.82 Å. In the absence of internal (H_2) motions this distance gives, via Equation (8.12), a minimal relaxation time, ^1H $T_{1\,min}(H_2)$, of 0.00157 s ($\nu_H = 200$ MHz) compared with 0.0040 s measured experimentally. It is obvious that, in the absence of knowledge of the internal dynamics, calculations of H−H distances in such systems will require significant corrections [5].

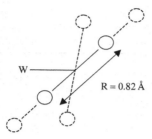

Figure 10.2 Dihydrogen ligand with ultra-fast vibrational and librational motions of the hydrogen atoms (R is an equilibrium internuclear distance; the Z-coordinate is along the H$-$H bond)

10.1 ^{1}H T_1 Times and H$-$H Distances in the Presence of Fast Vibrations and Librations

Ultra-fast stretching, bending and torsional motions are general properties of chemical bonds and are always present in investigated objects. Note however that the effects of these motions on calculated internuclear distances depend on the nature and conditions of the experiment. For this reason, the distances, determined by the solid-state NMR technique, by NMR relaxation in solutions and by neutron diffraction, are different.

Figure 10.2 shows vibrational and librational motions in the (H_2) ligand of the W dihydrogen complex with amplitudes Δ_Z, Δ_X and Δ_Y, respectively. In the presence of stretching motions, the orientation of the vector connecting two nuclei is unchanged. In contrast, librations reorient this vector. Under these conditions, an H$-$H distance (0.82 Å in the figure) becomes an *equilibrium internuclear separation*, marked as R. Owing to the nature of dipolar coupling, the vibrations and librations lead to observation of an *effective internuclear separation*, marked as R^{EFF}. Henry and Szabo [6] have analyzed the influence of vibrational/librational motions on internuclear distances determined via dipolar NMR relaxation or lineshapes in the solid-state NMR spectra. It has been established that the R^{EFF} distance is constantly *longer* than the equilibrium internuclear distance. According to Henry and Szabo, the R^{EFF} value can be expressed, as a function of R and Δ:

$$R^{EFF} \simeq R + [< \Delta_Z > - (2 < \Delta_Z^2 >)/R] + [(1/2R)(< \Delta_X^2 > + < \Delta_Y^2 >)] \tag{10.1}$$

The first term in the equation corresponds to the vibrational averaging of dipolar coupling. The second term reflects the librational effects. Variations in Δ_Z, Δ_X, and Δ_Y amplitudes in Equation (10.1) from 0.01 to 0.02 and 0.05 Å allow one to express quantitatively the influence of the motions on the effective internuclear distances R^{EFF} determined by relaxation measurements. The R^{EFF} values, shown in Table 10.2, correspond to ^{1}H $T_{1\,min}(R^{EFF})$ times of

Table 10.2 Effective H—H distances R^{EFF} and ^1H $T_{1\,min}$ times at 200 MHz for the dihydrogen ligand with equilibrium internuclear distance 0.820 Å in the presence of vibrational and librational motions of the hydrogen atoms

Δ_Z (Å)	$\Delta_X = \Delta_Y$ (Å)	R^{EFF} (Å)	$T_{1\,min}(R^{EFF})$ (s)	$[T_{1\,min}(R^{EFF}) - T_{1\,min}(R)]/$ $T_{1\,min}(R^{EFF})$ (%)
0.01	0.01	0.830	0.00169	7.0
0.02	0.02	0.839	0.00180	12.8
0.02	0	0.839	0.00180	12.8
0.05	0.05	0.867	0.0022	28.6
0.05	0	0.864	0.00215	26.9

0.00169, 0.00180 and 0.00215 s ($\nu_H = 200$ MHz) via Equation (8.12). In turn, at $R = 0.82$ Å, the ^1H $T_{1\,min}$ (R) time is calculated as 0.00157 s. As can be seen, the high-amplitude motions cause significant increases in ^1H $T_{1\,min}$ and the H—H distances will be *overestimated*. For example, at $\Delta_Z = \Delta_X = \Delta_Y = 0.05$ Å, the effective ^1H $T_{1\,min}$ time results in 0.867 Å, requiring a 5.4% correction for the vibrational/librational motions.

The problem of the corrections of internuclear distances, calculated from relaxation data, has a common character. However, quantitatively, the corrections will be different for different chemical bonds and molecules. For example, the influence of vibrations on the dipolar C—H coupling is small, and does not depend on the nature of molecules and their environments. It has been found that the vibrations in C—H bonds reduce the dipolar coupling constant by 3% and thus the effective C—H bond lengths increase by 1% only. In contrast, librational motions, reorienting the dipolar C—H vectors, are dependent on the nature of molecules. In common case, they can cause significant effective elongations of the C—H bonds up to 5.6% [6].

In addition, dihydrogen ligands are mobile, even in the solid state. Therefore, the neutron diffraction data collected in the solid state, also require corrections. However in contrast to nuclear relaxation, neutron diffraction *underestimates* H—H distances [7]. For example, the H—H distance of 0.82 Å, found in the neutron diffraction structure of dihydrogen complex $(CO)_3(PPr^i_3)_2$ $W(H_2)$, should be *longer* by few hundredths of an angstrom. In fact, the solid-state ^1H NMR signal of this complex has been fitted to the Pake doublet line with $r(H-H) = 0.890 \pm 0.006$ Å [8].

10.2 ^1H T_1 Times and H—H Distances in the Presence of Fast Rotational Diffusion

A C_{3v} rotation of CH_3 groups as well as a free rotation of dihydrogen ligands (rotational diffusion) around the axis in Figure 6.2, is well established by variable-temperature NMR spectra in solution and the solid-state [9]. In most cases, rotational diffusion occurs with rate constants $>10^5$–10^6 s^{-1} and thus it is a fast process on the NMR time scale. However, on the time scale of molecular tumbling, the process can be slow. In this case, the (H_2) rotation does not affect nuclear spin–lattice relaxation and such (H_2) ligands are considered as immobile ones.

When the rotational correlation time τ_{ROT} is much shorter than the correlation times of molecular tumbling ($\tau_{ROT} \ll \tau_{MOL}$) and activation energy of the rotation is smaller than that of molecular tumbling, then the (H_2) unit behaves as a free rotor with $\theta = 90°$ (Figure 6.2). In this case, the Woessner spectral density function [10], written for a $T_{1\,min}$ time, leads to equation:

$$r(H-H) = 4.611\ (T_{1\,min}/\nu)^{1/6} \qquad (10.2)$$

where r, $T_{1\,min}$ and ν are measured in Å, seconds and MHz. respectively. The coefficient, 4.611, is the result of a four-fold elongation of the $^1H\ T_1$ time in the presence of free rotation (compare with the value of 5.815 in Equation 8.12).

The decrease of dipolar coupling due to a fast rotation is a common phenomenon in nuclear relaxation. However, the effects are quantitatively different. For example, theoretically, a fast CH_3 diffusion (Figure 6.2) can lead to reducing the $^{13}C-^1H$ dipolar coupling by factor of 9. In this case the angle θ is equal to 109° and coefficient $(3\cos\theta^2 - 1)^2/4$ in Equation (6.6) is calculated as 0.11. That is why in most cases, $^{13}CH_3\ T_1$ times are quite long.

To show practically the problem of H−H distance determinations in mobile groups, consider Table 10.3 [11]. These data compare H−H bond

Table 10.3 H−H bond lengths (Å) in dihydrogen complexes determined in the solid state (SS) and in solutions from the $^1H\ T_{1\,min}$ times for fast-spinning (FS) and immobile (IM) dihydrogen ligands

Complex	$r(H-H)^{FS}$	$r(HH)^{IM}$	$r(H-H)^{SS}$
$Os(H_2)H(dppe)_2{}^+$	0.99	1.25	0.95
$Ru(H_2)H(dppe)_2{}^+$	0.885	1.12	0.89^c
trans-$Os(H_2)I(NH_4)^+$	1.207	1.52	1.21^c
$Mo(H_2)(CO)(dppe)_2$	0.99	1.25	0.735^a
			0.88^b
			0.85^c
$Cr(H_2)(CO)_3(PPr^i{}_3)_2$	0.78	0.99	0.86^b
			0.84^c
$W(H_2)(CO)_3(PPr^i{}_3)_2$	0.76	0.96	0.82^a
			0.89^b
			0.85^c
$Nb(H_2)(C_5H_4SiMe_3)_2(PMe_2Ph)^+$	0.93	1.17	1.17^c
trans-$HIr(H_2)Cl_2(PCy_3)_2$	0.74	0.94	0.85^c

^{a} Neutron diffraction distance; should corrected for libration to the range 0.80–0.88 Å
^{b} Solid-state ^1H NMR
^{c} Determined from the correlation $r(H-H) = 1.42 - 0.0167J_{HD}$

lengths in dihydrogen complexes, obtained from ^{1}H $T_{1\,min}$ times in solutions and diffraction methods in the solid-state. The relaxation data have been interpreted in terms of fast- and slow-spinning ligands. Undoubtedly, the first three complexes have dihydrogen ligands of the fast-spinning nature. In fact, a free rotation model gives, via Equation (10.2), H$-$H distances in very good agreement with the solid-state data.

According to the method of inelastic neutron scattering, the energy barrier of a (H$_2$) rotation in solid Mo(H$_2$)(CO)(dppe)$_2$ is as low as 0.7 kcal/mol. Reorientations of such molecules in solution require remarkably higher energies (see Table 1.4). Thus the ^{1}H relaxation in Mo(H$_2$)(CO)(dppe)$_2$ can be treated in the framework of a free rotation model. Then, the ^{1}H $T_{1\,min}$ time of 0.02 s, measured at 200 MHz, gives 0.99 Å compared with an unrealistically long distance (1.25 Å) expected for an immobile dihydrogen ligand (see Equation 8.12). Nevertheless, the H$-$H distance in the solid state (0.88 Å) is still shorter. It is probable that the value of 0.99 Å requires an additional correction for vibrational/librational (H$_2$) motions.

Rotation of the (H$_2$) ligand in solutions of complex Nb(H$_2$)(C$_5$H$_4$SiMe$_3$)$_2$(PMe$_2$Ph)$^+$ can be stopped on the NMR time scale at low temperatures [12]. Thus, the (H$_2$) rotation is slow on the time scale of molecular tumbling. Actually, the ^{1}H $T_{1\,min}$ time, measured in this complex as 0.020 s (300 MHz), gives 1.17 Å via Equation (8.12) in good agreement with structural data in the solid state.

The ^{1}H $T_{1\,min}$(H$_2$) times in complexes Cr(H$_2$)(CO)$_3$(PPr^i$_3$)$_2$ and *trans*-HIr(H$_2$)Cl$_2$(PCy$_3$)$_2$ lead to the unrealistically short H$-$H bond lengths, calculated in the approximation of a fast-spinning model [13]. Even without corrections for vibrational/librational motions, the H$-$H distances are very close to 0.75 Å in molecular hydrogen. Immobile (H$_2$) ligands (Equation 8.12) give more plausible distances (0.99 and 0.94 Å) to account for vibrational/librational (H$_2$) motions. All the examples demonstrate that determinations of H$-$H distances in rapidly moving molecular fragments require detailed information about the character and rates of internal motions. It is obvious that, in the absence of these details, uncertainties in r(H$-$H) calculations are quite large.

Comparison of the spectral density functions in Equation (6.6) and (4.5) shows that nuclear relaxation in immobile and mobile groups is not distinguishable phenomenologically. In the both cases, plots of the $\ln(T_1)$ versus $1/T$ are symmetric and V-shaped. The situation changes under the condition:

$$1/\tau(H_2) \simeq 1/\tau_{MOL} \approx \omega_H \qquad (10.3)$$

It is easy to show that in this case the plots of the $\ln(T_1)$ versus $1/T$ become *asymmetric* and sensitive to (H$_2$) motions [5, 14]. Similar effects have been observed for relaxation of ellipsoidal molecules in Chapter 6 (see Figure 6.2 at $\rho = 10$ and 50).

10.3 The Spectral Density Function for High-amplitude Librations

The solid-state ^{1}H NMR spectra of complex $W(H_2)(CO)_3(PPr^i_3)_2$ directly show (H_2) librations. The librations sweep out an average angle of $\pm 16°$, even at 100 K and below [4]. To describe nuclear relaxation in such systems quantitatively, Morris and Wittebort [5] have considered librations in a pair of protons as reorientations of the dipolar H–H vector in the limits of the angle amplitude, Φ, (see Figure 10.3) at a constant internuclear distance. If the librations are fast on the time scale of molecular tumbling (i.e. $\tau_C(lib)$ $\ll \tau_C(mol)$), then the angle Φ can be incorporated into the expression of the relaxation rate to give the spectral density function:

$$J(\omega_H) = 0.25\,(\tau_c(mol))/(1 + \omega_H^2\tau_c^2(mol)) +$$
$$+\,0.75\,(1 - 4 < \Phi^2 >)(\tau_c(mol)/(1 + \omega_H^2\tau_c^2(mol))$$
$$J(2\omega_H) = 0.25(\tau_c(mol)/(1 + 4\omega_H^2\tau_c^2(mol)) +$$
$$+\,0.75\,(1 - 4 < \Phi^2 >)(\tau_c(mol)/(1 + 4\omega_H^2\tau_c^2(mol)) \quad (10.4)$$

At $\Phi = 0°$ a dihydrogen ligand is immobile and Equation (10.4) coverts to the Bloembergen–Purcell–Pound spectral density function. According to Morris and Wittebort, in the presence of ultra-fast librations, H–H distances can be calculated via the minimal ^{1}H T_1 times by the equation:

$$r(H-H) = C(\Phi)(T_{1\,min}/v)^{1/6} \quad (10.5)$$

where $T_{1\,min}$ and v are measured in seconds and MHz, respectively. In turn, coefficients $C(\Phi)$, reflecting the influence of ultra-fast librations, are computed at variations in the Φ amplitudes (Table 10.4).

As we have shown, the H–H bond length calculations for the fast-spinning and immobile ligands in complex $W(H_2)(CO)_3(PPr^i_3)_2$ from the solution ^{1}H $T_{1\,min}$ time disagree with the solid-state data (Table 10.3). Take the model of a librating dihydrogen ligand and calculate the Φ. For an immobile (H_2)

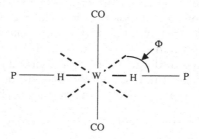

Figure 10.3 Ultra fast librations of the dihydrogen ligands in complex $W(H_2)(CO)_3(PPr^i_3)_2$. The librations are limited by the angle amplitude, Φ

Table 10.4 Librational corrections $C(\Phi)$ calculated on the
basis of Equation (10.4) at different amplitudes Φ [5]

Φ (°)	$C(\Phi)$
0	5.82
9.05	5.74
12.8	5.66
15.7	5.57
18.1	5.48
20.2	5.37
22.2	5.26
25.6	4.98
28.6	4.59
Fast-spinning	4.611

ligand, the ^1H $T_{1\,min}$ time of 0.004 s (200 MHz) gives 0.96 Å via Equation (8.12) compared with 0.89 Å, found in the solid state. Then, a simple combination of Equations (8.12) and (10.5) leads to $C(\Phi) = 5.401$. According to Table 10.4, the latter corresponds to an angle amplitude Φ, between 18 and 20°. Note that the solid-state ^1H NMR spectra of $W(H_2)(CO)_3(PPr^i_3)_2$ have shown (H_2) librations with amplitudes Φ of $\pm 16°$ at 100 K. A slight increase of Φ on going from the solid state to solutions is plausible, and thus the librating model is supported experimentally.

10.4 90° Jumps in a Four-fold Potential

By definition, the H—H vector in a fast-spinning (H_2) ligand can take any orientation in the plane perpendicular to the rotational axis. However, in many cases a dihydrogen ligand occupies two possible positions for energetic reasons (Figure 10.4). Usually, the energy barrier between the positions is quite low and thus the H—H vector undergoes fast 90° jumps in the four-fold potential. Note that spin populations in these states can be different.

Morris and Wittebort [5] have analyzed nuclear relaxation in such a system in terms of a superposition of two states, A and B (Figure 10.4). In both states, the H—H vector undergoes isotropic reorientations with molecular motion correlation times $\tau_c(mol)$. Under these conditions, the spectral density function, depending on rates of A/B transformations, is expressed by:

$$J(\omega_H) = (P_A^2 - P_A P_B + P_B^2)(\tau_c(mol)/(1 + \omega_H^2\tau_c^2(mol))$$
$$+ 3P_A P_B[\tau_c(mol) + (k_A + k_B)\tau_c^2(mol)]/$$
$$\{[1 + (k_A + k_B)\tau_c(mol)]^2 + \omega_H^2\tau_c^2(mol)\} \quad (10.6)$$

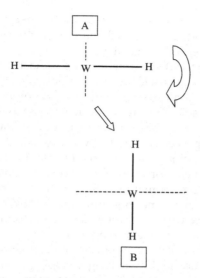

Figure 10.4 Schematic representation of 90° jumping dihydrogen ligands

where P_A and P_B are the populations of A and B states, $P_A + P_B = 1$ and k_A and k_B are the corresponding rate constants. If the k values are small (case a):

$$k_A + k_B \ll 1/\tau_c(\text{mol}) \qquad \text{(a)}$$

$$k_A + k_B \gg 1/\tau_c(\text{mol}) \qquad \text{(b)} \qquad (10.7)$$

then Equation (10.6) converts to the Bloembergen–Purcell–Pound spectral density function. In case b, the second term in Equation (10.6) transforms to zero and the function is simplified. Incorporating the function into the expression for the ^{1}H $T_{1\,\text{min}}$ time leads to:

$$r(\text{H–H}) = C(P)(T_{1\,\text{min}}/\nu)^{1/6} \qquad (10.8)$$

In turn, the $C(P)$ factor is:

$$C(P) = 5.815\,(P_A{}^2 - P_A P_B + P_B{}^2)^{1/6} \qquad (10.9)$$

One can show that 90° jumps and a free rotation model are identical at $P_A = P_B = 0.5$.

Consider application of this formalism for complex $W(H_2)(CO)_3(PPr^i{}_3)_2$ where ^{1}H $T_{1\,\text{min}} = 0.004$ s at 200 MHz. As we have shown, the H–H bond length of 0.96 Å is calculated from the $T_{1\,\text{min}}$ time for an immobile dihydrogen ligand compared with 0.89 Å, measured in the solid state. Then, a combination of Equations (8.12), (10.8) and (10.9) gives $P_A = 0.7$.

An analysis of Equation (10.6) shows that at $k_A + k_B \ll 1/\tau_c(\text{mol})$ the plots of $\ln(T_1)$ versus $1/T$ are symmetric and V-shaped. However if the equilibrium constant, $K = P_A/P_B$, depends strongly on the temperature, then the variable-temperature relaxation curves can deviate from the symmetrical V-shapes. This effect can be observed experimentally.

To conclude this part, it must again be emphasized that, in the absence of knowledge of internal molecular dynamics, accurate determinations of internuclear distances in flexible systems are not possible because commonly nuclear relaxation in solutions does not distinguish the types of motions. In this context, relaxation in the solid state has significant advantages. First, the temperature diapason is practically unrestricted. Second, the type of internal molecular motions can be directly established from the character of the relaxation curves. For example, a *tunneling* rotation of CH$_3$ groups is well identified as an anomalous ^{1}H T_1 behavior in the field-dependent experiments [15]. The experiments require special NMR techniques providing the field-cycling procedures: (a) saturation of proton magnetization with a comb of radiofrequency pulses at magnetic field B_{NMR}; (b) rapid switching to a new magnetic field B_R; (c) evolution of the magnetization in this field B_R; (d) rapid magnetic field switching to B_{NMR}; (e) measurements of the magnetization by pulses at field B_{NMR}. Some examples of the variable-field relaxation experiments on solutions will be considered in Chapter 11.

10.5 Deuterium Spin − Lattice NMR Relaxation in Mobile Molecular Fragments

Spin−lattice relaxation of deuterium is completely governed by quadrupole interactions and at isotropic molecular motions, the relaxation rate is:

$$1/T_1 = (3/10)\pi^2(DQCC)^2(1 + \eta^2/3)$$
$$\times (\tau_c/(1 + \omega_D^2\tau_c^2) + 4\tau_c/(1 + 4\omega_D^2\tau_c^2)) \tag{10.10}$$

where DQCC is the *static* deuterium quadrupole coupling constant and η is the asymmetry parameter of the electric field gradient at deuterium. The term 'static' is relative to DQCC values determined in static molecules. The V-shaped plots of $\ln(T_1)$ versus $1/T$ go through minima ($T_{1\,min}$) and then:

$$1/T_{1\,min} = 0.672\,(1 + \eta^2/3)(DQCC)^2/\nu_D \tag{10.11}$$

where $T_{1\,min}$, ν_D and DQCC are measured in seconds, MHz and kHz, respectively. Note that, in most chemical bonds, the η parameter is close to zero. Therefore, at the known static DQCC values, Equation (10.11), gives good predictions of ^{2}H $T_{1\,min}$ times, and vice versa, the ^{2}H $T_{1\,min}$ times provide determinations of the DQCC. The task of this part is to show how this expression changes in the presence of fast internal motions. Good models for these considerations are (D$_2$) ligands in dihydrogen complexes.

In fact, they can be classified as immobile, fast-spinning and 90° or 180° jumping ligands.

An immobile (D_2) ligand. Even in the case of immobile X–D bonds, the X and D atoms undergo ultra-fast vibrational/librational motions. According to the theory of quadrupole interactions, the electric field gradient at D depends on the X–D bond length (see Equation 4.22) and the principal eq_{ZZ} component is along this bond. Then, by analogy with nuclear relaxation in a pair of protons, the vibrations and librations in X–D bonds will partially average the quadrupolar coupling and thus DQCC values, determined by $^2H\ T_{1min}$ measurements, will be effectively reduced. Henry and Szabo have shown that this effect can reach 6% of the DQCC [6] and thus the measured magnitudes should be corrected.

180° Jumps of (D_2) ligands in solids. It has been established that dihydrogen ligands are mobile, even in the solid state and at low temperatures: hydrogen atoms can change their positions due to 180° jumps shown in Figure 10.5. It should be noted that orientation of the dipolar H–H vector in the (H_2) ligand, lying along the H–H bond, does not change due to this type of 180° jump. For this reason, this motion does not effect the $^1H\ T_1$ time. In contrast, the eq_{ZZ} component at D in a (D_2) ligand deviates from the D–D direction. Under these conditions the ^2H NMR relaxation rate $1/T_1(EX)$ is:

$$1/T_1(EX) = (9/160)\,(1 + \eta^2/3)\,(\sin 2\alpha)^2$$
$$(DQCC)^2(\tau_{EX}/(1 + \omega_D^2\tau_{EX}^2) + 4\tau_{EX}/(1 + 4\omega_D^2\tau_{EX}^2)) \qquad (10.12)$$

where 2α is the angle formed by principal eq_{ZZ} components at two deuterons and τ_{EX} is the correlation time of the exchange [16]. When the $^2H\ T_1$ time in Equation (10.12) reaches a minimum, it can be rewritten as [17]:

$$1/T_{1min}(EX) = 0.0128\,(1 + \eta^2/3)\,(\sin 2\alpha)^2\,(DQCC)^2/\nu_D \qquad (10.13)$$

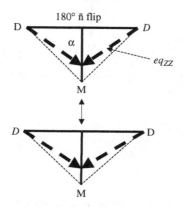

Figure 10.5　180°-Jumps of a (D_2) ligand around the axis perpendicular to the D–D bond

A free (D_2) rotation in solution. In solution, ^{2}H T_1 times depend on molecular tumbling (the molecular motion correlation times τ_C) and internal high-amplitude (D_2) motions. At slow internal motions on the ω_D time scale, dihydrogen ligands are immobile and their relaxation behavior corresponds to Equation (10.10) or (10.11). In fast-spinning dihydrogen ligands, D–D bonds undergo a free rotation (Figure 10.6), correlation times of which are significantly shorter than the τ_C. Nuclear relaxation in such free rotors is described by the Woessner spectral density function. For ^{2}H T_1 times, the Woessner equation gives:

$$1/T_{1\min}(\text{ROT}) = 0.168 \, (3\cos^2\alpha - 1)^2 \, (1 + \eta^2/3) \, (\text{DQCC})^2/\nu_D \qquad (10.14)$$

where α is the angle formed by the principal electric field gradient component, eq_{ZZ}, and the rotational axis.

(D_2) librations in solution. By analogy with librations of (H_2) ligands in a two-fold potential, fast (D_2) librations with amplitudes Φ (Figure 10.7) will affect ^{2}H $T_{1\min}$ times according to:

$$1/T_{1\min}(\text{LIB}) = 0.672 \, F(\alpha, \phi)(1 + \eta^2/3) \, (\text{DQCC})^2/\nu_D \qquad (10.15)$$

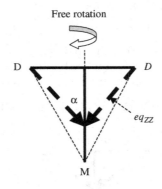

Figure 10.6 A dihydrogen (D_2) ligand undergoing free rotation around the axis, perpendicular to the D–D bond

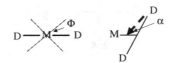

Figure 10.7 Ultra-fast (D_2) librations in a dihydrogen ligand. The librations are limited by the amplitude angle, Φ (α is the angle formed by the eq_{ZZ} vector and the motional axis)

where factors $F(\alpha, \phi)$ can be calculated at variations in the α and ϕ values [18] by analogy with the spectral density function suggested by Morris and Wittebort (see Equation 10.4).

180° jumps in solution. Topologically, 180° reorientations of the (D_2) ligand in Figure 10.5 correspond to displacement of the principal eq_{ZZ} component by the angle 2α. If fast 180° jumps occur in solution this (D_2) system can be regarded as two equally populated states, which undergo isotropic motions [19]. Then, the ^2H $T_{1\,min}$ can be written as:

$$1/T_{1\,min}(180°) = 0.672\, C_1\, (1 + \eta^2/3)\, (DQCC)^2/\nu_D \tag{10.16}$$

where $C_1 = 0.25(1 + 3\cos^2 2\alpha)$. It is seen that at $\alpha = 0$ Equation (10.16) transforms to Equation (10.11). Note that, according to Morris and Wittebort, 90° (H_2) jumps in a four-fold potential and a fast (H_2) rotation are equivalent when $H-H$ orientations do not have an energetic preference. Therefore, we do not deal with this case.

The Equations (10.11) to (10.16) show that fast internal motions increase ^2H $T_{1\,min}$ times. It is important that the resulting effect depends on the nature of motions and the DQCC, α and η values. It is obvious that the quadrupolar parameters cannot be accurately determined by the ^2H T_1 relaxation when the nature of internal motions is unknown. Therefore in most cases, ^2H T_1 experiments are focused on studies of molecular motions in solutions of ^2H-labeled compounds containing simple chemical bonds (for example, ^2H-C, ^2H-N). Since in this case, the deuterium quadrupole parameters are well established and the eq_{ZZ} components are lying along the chemical bonds, the T_1 values can be easy fitted to the corresponding motional model.

It can be demonstrated, however, that, even in the presence of many uncertainties, ^2H $T_{1\,min}$ data can still provide some useful conclusions. First, the uncertainty connected with the unknown η parameter seems to be insignificant because the $(1 + \eta^2/3)$ factor changes from 1 to 1.33 only. Therefore we take $\eta = 0$ as a good approximation. Second, when α in these equations approaches the magic angle (54°) and amplitudes ϕ (Figure 10.7) are high, then the rotational and librational motions have a maximal influence on the $^2H\,T_{1\,min}$ times. In the case of 180° jumps, the $^2H\,T_{1\,min}$ will be longest at $\alpha = 45°$ (see Equation 10.16). As the DQCC in Equations (10.11) and (10.14) to (10.16) remains constant, these mentioned conditions lead to the maximally possible $T_{1\,min}$ increases by a factor >4, by 4 and by 2.8 caused by the rotation, 180° jumps and librations, respectively. From the elongating factors it can be concluded that the dihydrogen ligands in dihydrogen/hydride complexes can be classified as fast-spinning ones if the $^2H\,T_{1\,min}$ increases by a factor >4 on going from D to (D_2). This conclusion is true because the DQCC values in (D_2) ligands are higher than in (D) ligands (see below) and this relaxation criterion can be used for structural formulations of hydride systems, even in the absence of additional data.

The complex $Ru(D_2)Cl(dppe)_2^+$ illustrates determination of internal (D_2) motions in solution, based on quadrupolar parameters, obtained by solid-state ^2H NMR. According to the solid-state ^2H NMR spectrum recorded at 5.4 K, the tensor of the electric field gradient in the solid complex $Ru(D_2)Cl(dppe)_2^+$ is axially symmetric ($\eta = 0.1$), its principal component, eq_{ZZ}, strongly deviates from the D–D bond ($\alpha = 45°$) and the *static* DQCC value is equal to 107 kHz [20]. The ^2H $T_{1\,min}$ time in the solid complex is measured as 0.161 s at 61.45 MHz (for convenience, all the T_1 times to be discussed are converted to the ^2H NMR frequency of 61.45 MHz). Then, the exact DQCC, α and η values lead to a ^2H $T_{1\,min}$ value of 0.42 s, expected for an 180° jump diffusion (Equation 10.13). The predicted time is longer and therefore the (D_2) motions are a combination of two-site jumps and a rotational diffusion. For example, they could be a combination of 180° jumps with 90° jumps in a four-fold potential. Owing to fast molecular tumbling, the ^2H $T_{1\,min}$ in solutions of this complex [21] is shorter than in the solid state (0.047 s and 0.161 s, respectively). Morris and Wittebort have classified the (H_2) ligand in $Ru(H_2)Cl(dppe)_2^+$ as a fast-spinning one on the basis of ^1H T_1 NMR data in solution. However, the solid-state quadrupolar parameters and a free rotation model, applied for a solution of $Ru(D_2)Cl(dppe)_2^+$, predict a long ^2H $T_{1\,min}$ time of 0.128 s against 0.047 s measured experimentally. In contrast, an 180° jump model, applied for a solution (Equation 10.16) leads to a shorter ^2H $T_{1\,min}$ time (0.032 s). Thus, internal (D_2) motions in solutions of $Ru(D_2)Cl(dppe)_2^+$ are again a combination of rotational diffusion with 180° jumps (probably in a ratio of 0.4/0.6).

The hydride and dihydrogen ligands in solutions of the complex $RuD(D_2)(dppe)_2^+$ show ^2H $T_{1\,min}$ times of 0.015 and 0.181 s, respectively [21]. Even in the absence of exact quadrupolar parameters, these data reveal the fast-spinning nature of the (D_2) ligand to be in good agreement with formulation based on the ^1H T_1 NMR relaxation [5]. It has been established that (D_2) motions in Ru dihydrogen complexes can be probed by *static* DQCC magnitudes taken from DFT calculations and corrected as 107–117 kHz [17]. Then, orientation of the eq_{ZZ} component at D in the dihydrogen ligand of $RuD(D_2)(dppe)_2^+$ is calculated via Equation (10.14) as $\alpha = 46–48°$.

Solutions of dihydrogen complex $Ru(D_2)Cp^*(dppm)^+$ show ^2H $T_{1\,min}$ of 0.019–0.021 s. According to the ^1H T_1 NMR relaxation data [5], ^1H iso-topomeric complex $Ru(H_2)Cp^*(dppm)^+$ is one of the innumerous examples, containing an immobile (H_2) ligand on the time scale of molecular tumbling. Table 10.5 lists the ^2H $T_{1\,min}$ times calculated at DQCC = 107–117 kHz (taken from DFT computations of Ru dihydrogen systems) for the static, 180° jumping and librating (D_2) ligands. It is obvious that the (D_2) ligand is not static in $Ru(D_2)Cp^*(dppm)^+$ at the deuterium frequency. That is why the deuterium $T_{1\,min}$ time (0.014–0.018 s) in the classical dihydride isomer, $RuD_2Cp^*(dppm)^+$, is shorter than that in the dihydrogen isomer. Since 180°

Table 10.5 ^{2}H $T_{1\,min}$ times (61.45 MHz) calculated for static ($T_{1\,min}$) and moving dihydrogen ligands at different DQCC values

Complex	$T_{1\,min}$(DQCC)	$T_{1\,min}180°$(DQCC)[a]	$T_{1\,min}$LIB(DQCC)[b]
Ru(D$_2$)	0.00799 (107)	0.0320 (107)	0.0224 (107)
	0.00645 (119)	0.0258 (119)	0.0180 (119)

[a] Calculated as a maximally long $T_{1\,min}$ time at $\alpha = 45°$
[b] Calculated as a maximally long $T_{1\,min}$ time at $\phi = 23°$ and $\alpha = 54°$

Table 10.6 Geometries and orientations of the principal components the EFG tensors at D in the dihydrogen ligands of some dihydrogen complexes

Compound	r(D–D) (Å)	r(M–D) (Å)	DMD (°)	2α (°)	θ (°)
W(D$_2$)(CO)$_3$(PCy$_3$)$_2$	0.89	1.9	27	68	104
RuD(D$_2$)(dppe)$_2$$^+$	0.94	1.81	30	92–96	105
OsD(D$_2$)(dppe)$_2$$^+$	0.96	1.74	32	80–84	106
Ru(D$_2$)Cp*(dppm)$^+$	1.1	1.66	38	64–72	109
Os(D$_2$)Cl(dppe)$_2$$^+$	1.22	1.58	45	40–66	113
				40–72	

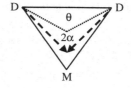

Figure 10.8 Orientation of the principal components of the electric field gradients at D (dashed arrows) determined for dihydrogen ligands by the ^{2}H T_1 relaxation measurements in account for fast internal (D_2) motions (angles θ and 2α are formed by two MDD bisectors and eq_{ZZ} vectors, respectively)

jumps do not affect ^{1}H T_1 times, they contribute to the deuterium relaxation rate in Ru(D$_2$)Cp*(dppm)$^+$. Then, Equation (10.16) results in α values in the range 32–36°. Thus in spite of uncertainties connected with the high internal mobility of dihydrogen ligands, ^{2}H NMR relaxation leads to an important chemical conclusion: the principal components of the electric field gradients at D in dihydrogen ligands are oriented closer to the M–D directions (see Table 10.6 and Figure 10.8) illustrating large contributions of metal atoms to (H$_2$) bonding. Finally, it must be emphasized that the data discussed illustrate a higher sensitivity of deuterium relaxation to motions with respect to proton relaxation.

Bibliography

1. P. J. Desrosiers, L. Cai, Z. Lin, R. Richards, J. Halpern. *Journal of the American Chemical Society* 1991; **113**: 4173.

2. H. H. Limbach, G. Scherer, M. Maurer, B. Chaudret. *Angewanote Chemie International Edition in English* 1992; **31**: 1369.

3. E. Clot, C. Leforestier, O. Eisenstein, M. Pelissier. *Journal of the American Chemical Society* 1995; **117**: 1797.

4. G. J. Kubas. *Metal Dihydrogen and σ-Bond Complexes.* Kluwer/Plenum: New York, 2001.

5. R. H. Morris, R. Wittebort. *Journal of Magnetic Resonance Chemistry* 1997; **35**: 243.

6. E. R. Henry, A. Szabo. *Journal of Chemical Physics* 1985; **82**: 4653.

7. G. J. Kubas, J. E. Nelson, J. C. Bryan, J. Eckert, L. Wisniewski, K. W. Zilm. *Inorganic Chemistry* 1994; **33**: 2954.

8. K. W. Zilm, R. A. Merrill, M. W. Kummer, G. J. Kubas. *Journal of the American Chemical Society* 1986; **108**: 7837.

9. D. M. Heinekey, W. J. Oldham. *Chemical Reviews* 1993; **93**: 913.

10. D. E. Woessner. *Journal of Chemical Physics* 1962; **36**: 1.

11. P. A. Maltby, M. Schlaf, M. Steinbeck, A. J. Lough, R. H. Morris, W. T. Klooster, T. F. Koetzle, R. C. Srivastava. *Journal of the American Chemical Society* 1996; **118**: 5396.

12. F. A. Jalon, A. Otero, B. R. Manzano, E. Villasenor, B. Chaudret. *Journal of the American Chemical Society* 1995; **117**: 10123.

13. D. G. Gusev, R. L. Kuhlman, K. B. Renkema, O. Eisenstein, K. G. Caulton. *Inorganic Chemistry* 1996; **35**: 6775.

14. W. T. Klooster, T. F. Koetzle, G. Jia, T. P. Fong, R. H. Morris, A. Albinati. *Journal of the American Chemical Society* 1994; **116**: 7677.

15. A. J. Horsewill. *Progress in Nuclear Magnetic Resonance Spectroscopy* 1999; **35**: 359.

16. F. Wehrmann, J. Albrecht, E. Gedat, G. J. Kubas, J. Eckert, H. H. Limbach, G. Buntkowsky. *Journal of Physical Chemistry A* 2002; **106**: 1977.

17. V. I. Bakhmutov. *Magnetic Resonance in Chemistry* 2004; **42**: 66.

18. V. I. Bakhmutov, C. Bianchini, F. Maseras, A. Lledos, M. Peruzzini, E. V. Vorontsov. *Chemistry. A European Journal* 1999; **5**: 3318.

19. L. Latanowicz. *Berichte den Bunsengesellscheft für Physitcalische Chemie* 1987; **91**: 237.

20. F. Wehrmann, T. P Fong, R. H. Morris, H. H. Limbach, G. Buntkowsky. *Physical Chemistry and Chemical Physics* 1999; **1**: 4033.

21. G. A. Facey, T. P. Fong, D. G. Gusev, P. M. Macdonalds, R. H. Morris, M. Schlaf, W. Xu. *Canadian Journal of Chemistry* 1999; **77**: 1899.

11 Relaxation of Nuclei Other Than ^{1}H and ^{2}H and Specific Relaxation Experiments

By analogy with protons and deuterons, relaxation of other nuclei can be successfully used to solve the following problems: experimental determinations of chemical shift anisotropies and quadrupole coupling constants (QCC), when direct methods (solid state NMR and NQR, respectively) are not available; investigations of weak interactions such as complexation and association or hydrogen bonding; identifications and studies of chemical exchanges and molecular motions. Methodologies of these investigations are quite different and depend on the nature of nuclei and relaxation mechanisms. This chapter includes multinuclear direct and indirect relaxation measurements, determinations and interpretations of cross-correlation relaxation rates, combinations of T_1 and T_2 experiments and the variable-field

Practical NMR Relaxation for Chemists   Vladimir I. Bakhmutov
© 2004 John Wiley & Sons, Ltd   ISBNs: 0-470-09445-1 (HB); 0-470-09446-X (PB)

relaxation measurements. The best way to introduce this topic is to show typical examples involving nuclei of different nature. Among them, ^{205}Tl, ^{103}Rh, ^{119}Hg and ^{59}Co seem to be important in organometallic chemistry, ^{23}Na and ^{27}Al illustrate relaxation studies of association and complexation and ^{14}N, ^{15}N and ^{13}C are often used as target nuclei in investigations of biomolecules.

11.1 Chemical Shift Anisotropies and Nuclear Quadrupole Coupling Constants from T_1 Times of Heavy Nuclei in Solution

In spite of the fact that chemical shift anisotropies $\Delta\sigma$ characterize magnetic properties of nuclei, the $\Delta\sigma$ values depend on local environments of nuclei and connect with the symmetry of compounds. Thus, they can provide important chemical information. Since the chemical shift anisotropy affects lineshapes in NMR spectra of solid static samples, solid-state NMR is a direct way to CSA determinations. If however, for some reason, the solid-state NMR spectra are unavailable, nuclear relaxation in solution becomes a unique method for CSA studies.

The CSA mechanism is dominant in relaxation of heavy nuclei such as ^{205}Tl, ^{195}Pt, ^{207}Pb, ^{57}Fe and ^{103}Rh. In spite of this dominance, CSA contributions should correctly evaluated from total relaxation rates because dipole–dipole and/or spin–rotation interactions can be also significant. At fast molecular motions (see Equation 5.6 at $1 \gg \omega^2\tau_c^2$), the rate of CSA relaxation is:

$$(T_1)^{-1}{}_{CSA} = (2/15)\, \gamma_I^2\, B_0^2\, (\Delta\sigma)^2\, \tau_c \tag{11.1}$$

If the total relaxation rate $(T_1)^{-1}{}_{OBS}$ increases linearly with the square of the applied magnetic field B_0 in variable-field experiments, then the CSA contribution $(T_1)^{-1}{}_{CSA}$, can be evaluated via the proportionality

$$(T_1)^{-1}{}_{(CSA)}(B^{(1)}) = [(T_1)^{-1}{}_{OBS}(B^{(1)}) - (T_1)^{-1}{}_{OBS}(B^{(2)})]$$
$$\times\, [1 - (B^{(2)})/(B^{(1)})^2]^{-1} \tag{11.2}$$

where $B^{(1)}$ and $B^{(2)}$ are the strengths of the applied external magnetic field. This feature is observed, for example, in relaxation of ^{103}Rh nuclei in complex Rh(COD)(tetrakis(1-pyrazole)borate) [1]. It is easy to show that the plot of the $(1/T_{1OBS})$ versus B_0^2 (see Table 11.1) is actually linear and has a *small intercept*. The latter is a good evidence for the negligibly small contributions from other mechanisms: CSA contributions to the total ^{103}Rh T_1 relaxation rates are calculated as 92 and 97% at magnetic fields of 7.05 and 11.75 T, respectively. It is obvious that a $\Delta\sigma(^{103}Rh)$ value can be calculated via

Table 11.1 [103]Rh T_1 relaxation time in a $CDCl_3$ solution of Rh(COD)(tetrakis(1-pyrazole)borate) measured as a function of the applied magnetic field strength B_0 at 298 K

B_0 (T)	[103]Rh T_1(OBS) (s)	$B_0{}^2(T^2)$	$1/T_1$(OBS) (s^{-1})
7.05	0.886	49.7	1.13
9.39	0.510	88.17	1.96
11.75	0.334	139.06	2.99

Equation (11.1) if the correlation time τ_C is known. Since olefin carbons in the Rh complex relax completely by dipole–dipole C–H interactions, their relaxation rate is:

$$1/T_1(\text{C–H}) = N \, (\mu_0/4\pi)^2 \, \gamma_H{}^2 \, \gamma_C{}^2 \, \hbar^2 \, r(\text{C–H})^{-6} \, \tau_C \qquad (11.3)$$

where N the number of hydrogens attached to the carbons. Then the [13]C T_1 time, measured for these carbons, and C–H distances of 1.09 Å give $\tau_C = 5.2 \times 10^{-11}$ s. The latter finally leads to $\Delta\sigma(^{103}\text{Rh}) = 6500$ ppm [1].

[205]Tl T_1 times in solutions of thallium compounds are shown in Table 11.2 [2]. The [205]Tl relaxation depends strongly on the applied magnetic field in the first four thallium compounds. The CSA mechanism is dominant and thus $\Delta\sigma$ values are determinable (310–1300 ppm). In the case of other compounds, the variable-field effects on T_1 are insignificant. [205]Tl nuclei relax simultaneously by the spin–rotation and CSA mechanisms. For this reason, $\Delta\sigma$ determinations are not possible.

In the presence of pronounced dipole–dipole interactions, CSA contributions can be accurately evaluated from total relaxation rates by partial deuteration of the samples or NOE measurements in combination with variable-field experiments. Spin–rotation contributions can be estimated if spin–rotation constants are known. Multinuclear NMR experiments on 3,5-dichlorophenylmercury cyanide in DMSO solutions illustrate the methodology, based on NOE measurements. It follows from the data in

Table 11.2 [205]Tl NMR spin–lattice relaxation times T_1 (s) in acidic water solutions ($HClO_4$) of compounds $Tl(X)_n(H_2O)_{m-n}{}^{(3-n)+}$ at different magnetic fields

Compound	115.4 MHz	230.8 MHz	288.5 MHz	Intercept (s^{-1})	$\Delta\sigma$(ppm)
$TiCl^{2+}$	0.19	0.06		1.4	1300
$TlCl_2{}^+$	0.18	0.042		−0.52	1600
$TlCl_3$	0.48	0.116		−0.096	960
$TlCl_4{}^-$	0.57	0.40		1.51	310
$TlBr_2{}^+$	0.39	0.27			
$TlBr_3$	0.93	1.08			
$TlBr_4{}^-$	0.80	1.20	1.28		

Table 11.3 Relaxation times (T_1, s), the NOE enhancements (η) and chemical shift anisotropies, $\Delta\sigma$, measured for ^{13}C, ^{15}N and ^{199}Hg nuclei, in 3,5-dichlorophenylmercury cyanide with ^{13}C and ^{15}N labels in the CN group (DMSO-d$_6$)

	Cpara	Cmeta	Cortho	^{13}C$_{CN}$	^{15}N$_{CN}$	^{199}Hg
T_1(4.7 T)	0.42	9	0.76	4.1	11.5	0.06
T_1(11.7 T)	0.42		0.78	0.75	2.3	0.01
η(4.7 T)	1.75	0.2	1.75	0		
$\Delta\sigma$ ppm				360 ± 30	500 ± 40	3245 ± 260
				$(338.2)^a$	$(637)^a$	

^{a}Values calculated at the DFT(B3LYP) level

Table 11.3 that the carbon resonances in the phenyl ring show large positive NOE enhancements (1.75), close to a maximal value ($\eta = 1.99$). In addition, the ^{13}C T_1 values are independent of the applied magnetic field. Thus, the ^{13}C relaxation of the phenyl carbons is completely governed by dipole–dipole carbon–proton interactions. In contrast, the ^{15}N and ^{13}C T_1 times in the CN group and also the ^{199}Hg T_1 values are field dependent. At the same time, the NOEs are invisible. It is obvious that the CSA mechanism is dominant in the relaxation of all the above nuclei and their T_1 times give the $\Delta\sigma(^{13}$C), $\Delta\sigma(^{15}$N) and $\Delta\sigma(^{199}$Hg) values in good agreements with DFT(B3LYP) calculations.

Experiments on the cobalt cluster shown in Figure 11.1 illustrate NQCC determinations for heavy nuclei by T_1 measurements in solution [3]. The complex has a C$_{3v}$ structure with nonequivalent apical and basal cobalt atoms and shows two signals in the solid-state ^{59}Co NMR spectrum. Linewidths of the resonances are estimated as 50 and 350 kHz. A two-pulse Hahn-echo experiment allows one to determine quadrupolar parameters (QCC = 12.4 MHz

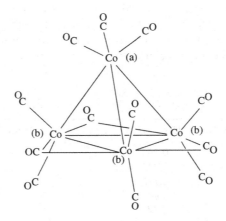

Figure 11.1 The structure of the cobalt cluster with the magnetically non-equivalent apical and basal cobalt atoms (Co(a) and Co(b), respectively)

Table 11.4 Variable-temperature ^{59}Co spin–lattice relaxation times (ms) determined for apical and basal Co atoms in the cluster $Co_4(CO)_{12}$ at ^{59}Co NMR frequency 71.21 MHz (CDCl$_3$)

T (K)	Apical ($\delta = -670$ ppm)	Basal ($\delta = -2030$ ppm)
291	1300	21.8
302	1560	26.8
312	1820	28.6
320	2040	35.0

and $\eta_{Co} = 0.32$) for the narrower signal, assigned to apical cobalt atoms. Because of the big broadening, no quadrupolar parameters can be obtained for basal Co atoms. In contrast to the solid state, both signals are well detected in ^{59}Co NMR spectra in a CDCl$_3$ solution and their T_1 times are accurately measured (Table 11.4) (note however that, owing to an apical/basal exchange, occurring at $T > 260$ K, the ^{59}Co relaxation is not exponential and the inversion recovery data require special treatment [3]). The relaxation rate for ^{59}Co nuclei ($I = 7/2$) can be expressed via Equation (11.4),

$$1/T_1(^{59}\text{Co}) = (2\pi^2/49)\, C_{QCC}\, \tau_C \qquad (11.4)$$

where the constant C_{QCC} is defined as $QCC^2(1 + \eta^2/3)$. In an isotropic approximation, the motion correlation times, τ_C, are identical for the apical and basal nuclei. Then Equation (11.5), combined with the solid-state NMR data for the narrow line and the solution ^{59}Co T_1 measurements, leads to $QCC_{basal} = 100$ MHz.

$$C_{QCC}(\text{basal}) = C_{QCC}(\text{apical})\, \{T_1(\text{apical})/T_1(\text{basal})\} \qquad (11.5)$$

11.2 Multinuclear Relaxation Approaches to Complexation, Association and H-bonding

As we have shown in the previous chapters, nuclear relaxation is very sensitive to molecular motions. Under extreme narrowing conditions (i.e. at $1 \gg \omega^2\tau_c^2$), the plots of $\ln(T_1)$ versus $1/T$ are linear and their slopes correspond to activation energies E_a of molecular motions. When individual components interact, a complex formed moves more slowly through the medium, requiring larger activation energy. The latter, being an energetic measure of intermolecular interactions, is observed by variable-temperature relaxation experiments. If the interactions lead to complexes of two or more types, plots of $\ln(T_1)$ versus $1/T$ can show several distinct linear regions. This

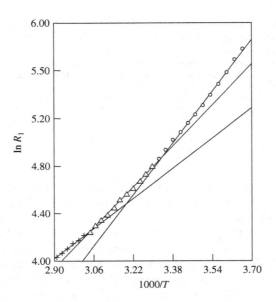

Figure 11.2 Variable-temperature relaxation rates of ^{23}Na nuclei, measured in a molten mixture of NaCl, AlCl$_3$ and 1-methyl-3-ethylimidazolium sodium chloride (Figure 11.3): Na $_{0.22}$ MEI $_{0.76}$ AlCl$_4$. (Reproduced from W. R. Carper, J. L. Pflug, J. S. Wilkes. *Inorganica Chimica Acta* 1992; **193**: 201, with permission from Elsevier)

situation is illustrated in Figure 11.2 where relaxation of ^{23}Na nuclei shows three linear sections as good evidence for complexation.

One of effective approaches to studies of association in solutions is the so-called multiple spin probe method. The method consists of the standard variable-temperature T_1 measurements for target nuclei, belonging to different individual components. If one of the nuclei relaxes by the dipole–dipole mechanism (DD), for example ^{13}C, and the other nucleus is quadrupolar (Q), then relaxation rates R_1(DD) and R_1(Q) can be expressed by:

$$R_1(DD)/a = R_1(Q)/b \tag{11.6}$$

where $a = N\,(\mu_0/4\pi)^2\,\gamma_H^2\,\gamma_C^2\,\hbar^2\,r(C–H)^{-6}$ and $b = 0.3\pi^2\,[(2I+1)/I^2\,(2I-1)]\,(QCC)^2(1+\eta^2/3)$. It is obvious that Equation (11.6) is valid when the correlation times τ_C of the both target nuclei are identical or, in other words, the components form a complex. If this statement is true, then plotting the R_1(DD) rates versus the R_1(Q) rates will lead to a linear dependence with zero intercept. This feature is observed for relaxation of ^{13}C and ^{23}Na nuclei in molten salts, containing NaCl, 1-methyl-3-ethylimidazolium sodium chloride (MEICl) (Figure 11.3) and AlCl$_3$ [4]. On heating from 0° to 70°C, the ^{13}C T_1(DD) times, measured for carbons C(4,5) and C(2) in MEICl, reduce from 5.50 to 0.55 s, correlating well with the ^{23}Na T_1(Q) times (changing from 0.017

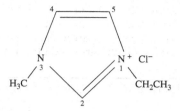

Figure 11.3 The molecular structure of 1-methyl-3-ethylimidazolium sodium chloride (MEICI)

to 0.003 s) obtained for NaCl. The complexation is independently supported by ^{23}Na R_1 and ^{27}Al R_1 measurements. Finally it should be emphasized that molecular volumes of the complexes formed can be characterized in terms of the Stokes radii by combination of T_1 experiments with viscosity measurements.

A typical hydrogen bond is shown in Figure 11.4 where B is the proton-accepting center and the α angle is close to 180°. The B centers usually contain quadrupolar nuclei (^{14}N or ^{17}O), relaxation of which is sensitive to hydrogen bonding. Methodologically relaxation studies of H-bonded systems can be considered by using pyridine/phenol mixtures in CCl$_4$ or CHCl$_3$ as examples [5]. Preliminary experiments, carried out on the individual pyridine in CCl$_4$, show a very small increase of the ^{14}N T_1 time from 1.59 to 1.64 ms at decreasing the pyridine concentration from 1.0 to 0.05 M. The effect certainly connects with a decrease of the solution viscosity. In contrast, an addition of phenol to a 0.5 M CCl$_4$ solution of pyridine is accompanied by a strong decrease of the ^{14}N T_1 time from 1.59 to 0.23 ms (Table 11.5). This effect is direct evidence for H-bonding. In fact, similar experiments, carried out on ^{14}N nuclei of pyridine in the presence of anisole or 2,6-di-*tert*-butylphenol, did not reveal the pronounced ^{14}N T_1 effects. The former cannot form H-bonds while the second compound has sterically hindered OH groups. In addition, the presence of phenol in solutions of benzaldehyde and acetonitrile (weaker proton acceptors than pyridine) does not perturb the ^{17}O, ^{13}C and ^{14}N T_1 times (Table 11.5).

Quantum-chemical calculations at the HF/6-311G(d,p) level predict a 20% *decrease* of the nitrogen quadrupole coupling constant in pyridine on going to

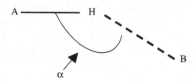

Figure 11.4 Schematic representation of a typical hydrogen bond where B is the proton-accepting center

Table 11.5 Room-temperature ^{14}N, ^{17}O and ^{13}C T_1 times in the individual proton acceptors and the H-bonded systems in solutions. The η values (NOE) are determined by ^{13}C and $^{13}C\{^1H\}$ NMR experiments. X_{HB} are molar fractions of H-bonded complexes found by IR spectra (reproduced with permission from A. Bagno *et al. Chemistry. A European Journal*. 2000; 6: 2915

Acceptor/solvent	^{14}N T_1 (ms)	^{17}O T_1 (ms)	^{13}C T_1 (s)	^{13}C T_1(DD) (s)	η
Pyridine/CCl$_4$	1.59		18.6 (C-4)	30.8 (C-4)	1.2
+ Phenol; X_{HB} = 0.82	0.23		2.6 (C-4)	3.1 (C-4)	1.7
Pyridine/CHCl$_3$	1.05		12.8 (C-4)	17.0 (C-4)	1.5
+ Phenol; X_{HB} = 0.82	0.35		3.9 (C-4)	4.3 (C-4)	1.8
Acetonitrile/CCl$_4$	2.43		34.0 (CN)	450 (CN)	0.15
+ Phenol; X_{HB} = 0.79	0.82		18.4 (CN)	229 (CN)	0.16
Benzaldehyde/CCl$_4$		1.95	16.2 (CO)	21.2 (CO)	1.7
+ Phenol; X_{HB} = 0.76		1.58	9.8 (CO)	11.7 (CO)	1.7

complex PhOH. . . NC$_5$H$_5$. Since the $1/T_1(Q)$ rate is proportional to the square of the quadrupole constant (see Equation 4.23), the ^{14}N T_1 time expected for this complex, should be longer than that in free pyridine. The observed effect is opposite and hence molecular dynamics prevails over the QCC change. For the same reason, the total and dipolar ^{13}C T_1 times, measured for carbons C(4) in pyridine, decrease strongly in the presence of phenol (Table 11.5).

Experiments on ^{15}N–substituted mono- and binuclear Pt amine cations $[Pt(^{15}NH_3)_4]^{2+}$ and $[\{Pt(^{15}NH_3)_3\}_2NH_2(CH_2)_4NH_2]^{4+}$ (Figure 11.5) illustrate how ^{15}N T_1 times change at cation/anion interactions between the Pt complexes with nucleotide 5′-guanosine monophosphate (5′-GMP) [5]. The data in Table 11.6 show a consistent decrease of the ^{15}N T_1 times when the concentration of 5′-GMP increases. As can be seen, the binuclear Pt complex demonstrates the bigger T_1 effects. Finally, note that these interactions can serve as a model of the approach of Pt-amine species to DNA.

11.3 ^{23}Na Relaxation in Solutions of Complex Molecular Systems

In contrast to small organic or inorganic molecules, complex molecular systems (for example, biological or polymer compounds) often show *slow* and *fast* reorientational motions on the NMR frequency scale. According to a two-step relaxation model, such motions are responsible for appearance of slow (S) and fast (F) relaxing components in NMR decays, collected experimentally. Under these conditions, the transverse and longitudinal magnetizations can be expressed by equations:

$$M_X = M_X^0[a\exp(-t/T_{2F}) + b\exp(-t/T_{2S})]$$

$$(M_Z - M_Z^0) = -M_Z^0[c\exp(-t/T_{1S}) + d\exp(-t/T_{1F})] \qquad (11.7)$$

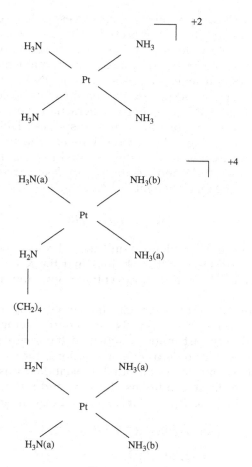

Figure 11.5 Structures of cationic ^{15}N–substituted mono- and binuclear Pt amine complexes, $[Pt(^{15}NH_3)_4]^{2+}$ and $[\{Pt(^{15}NH_3)_3\}_2NH_2(CH_2)_4NH_2]^{4+}$

Table 11.6 ^{15}N T_1 times (s), measured for platinum complexes in the presence of the mononucleotide 5'-GMP (guanosine-5'-monophosphate) in buffered aqueous solutions at 25°C

[Pt]/[5'-GMP]	Pt{(^{15}NH$_3$)$_4$]$^{2+}$ T_1	[{Pt(^{15}NH$_3$)$_3$}$_2$NH$_2$(CH$_2$)$_4$NH$_2$]$^{4+}$ T_1(N(a))	T_1(N(b))
	41.3	18.4	14.9
4/1	41.3	17.4	11.9
2/1	37.6	15.3	11.6
1/1	34.0	13.5	10.6
1/2	28.5	10.0	8.1
1/4	22.3	8.0	6.6

where a, b, c and d are portions of the corresponding components and therefore $a + b = 1$ and $c + d = 1$. It is obvious that treatments of these NMR decays require procedures for separation of two superimposed exponentials (the so-called F and S separations) by a convenient mathematical method.

Relaxation of ^{23}Na nuclei in water solutions of DNA is a typical example of the two-step behavior and illustrates well the methodology of relaxation applications for this field [7]. Owing to interactions between anionic sites of DNA and sodium ions (or other cationic species), DNA molecules are stabilized in solution. Thus, the sodium ions exist in bound (B) and free (F) states. This statement is basic to the so-called two-site relaxation model. An exchange between the B and F states is very fast on the NMR time scale and for this reason, the relaxation rates, ^{23}Na R_{OBS}, measured experimentally, are averaged:

$$R_{OBS} = P_F R_F + P_B R_F \qquad (11.8)$$

where P and R are the populations and the relaxation rates in the corresponding states, respectively. It is obvious that the R_{OBS} measurements at variations in DNA/^{23}Na ratios can eventually give the number of bound sodium ions.

In the framework of this two-step relaxation model, applied for DNA/^{23}Na systems, quadrupole interactions at ^{23}Na are rapidly averaged to a small (but nonzero) value due to fast motions and then the quadrupole coupling is effectively averaged out to zero by slow motions, for example, segmental motions within DNA molecules or radial translation diffusion of the bound sodium ions. Under these conditions, the slow and fast components of the ^{23}Na NMR relaxation, R_{1S}, R_{1F}, R_{2S} and R_{2F}, can be expressed as:

$$R_{1S} = (1/20)(e^2qQ/\hbar)^2 \, J(2\omega_0)$$

$$R_{1F} = (1/20)(e^2qQ/\hbar)^2 \, J(\omega_0)$$

$$R_{2S} = (1/40)(e^2qQ/\hbar)^2 \, [J(\omega_0) + J(2\omega_0)]$$

$$R_{2F} = (1/40)(e^2qQ/\hbar)^2 \, [J(0) + J(2\omega_0)] \qquad (11.9)$$

The spectral density functions J in Equations (11.9) have their usual meaning (see Section 8.3). Therefore, at relatively low magnetic fields, all the molecular motions can correspond to the extreme narrowing limit (i.e. $1 \gg \omega^2\tau_c^2$) and then the ^{23}Na relaxation is monoexponential and the R_{OBS} rates show only fast components. In this situation, a study of the association in ^{23}Na/DNA water solutions is strongly simplified. On the other hand, the low-field relaxation data provide no information about the dynamics of DNA molecules. In contrast, at high magnetic fields, the molecular motion correlation times τ_C, affecting the ^{23}Na relaxation, are not short compared with the inverse Larmor frequency. One can show that at $\omega_0\tau_C > 1.5$, the slow and fast components of spin–spin relaxation rates, $1/T_{2F}$ and $1/T_{2S}$, differ sufficiently

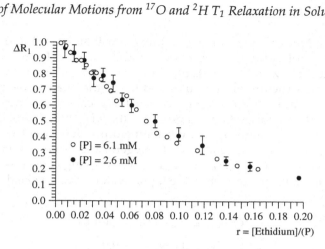

Figure 11.6 Decrease of the normalized enhancement of the ^{23}Na relaxation rate upon addition of ethilidium bromide in 2.6 and 6.2 mM NaDNA solutions. (Reproduced with permission from M. Case *et al.* *Biophysical Chemistry* 1996; **59**: 133. © John Wiley & Sons, Ltd.)

to be accurately separated, for example, by deconvolution of lineshapes observed in the ^{23}Na NMR spectra. Then the R_{2F} and R_{2S} values give R_{1S} and R_{1F}, respectively. Thus, the experiments at high magnetic fields are more informative. Methodologically, the studies of DNA/^{23}Na systems consist of titration of DNA with NaCl or NaDNA with multivalent competitors. Figure 11.6 illustrates an example of such a titration where the rate of the ^{23}Na spin–lattice relaxation in NaDNA molecules undergoes a ten-fold decrease at additions of ethidium bromide [8]. It is obvious that the observed effects directly connect with the appearance of free sodium ions which can be determined quantatively.

11.4 Character of Molecular Motions from ^{17}O and ^2H T_1 Relaxation in Solution

Quantitative characterization of molecular motions with the help of relaxation of quadrupolar nuclei is not a simple task, particularly when quadrupolar parameters at target nuclei are unknown or they can change with temperature. The simplest example is liquid water, forming H-bonds.

The motion correlation times of water molecules could be determined by ^2H and ^{17}O T_1 times via the standard equations. According to the data of Farrar *et al.* [9], the ^2H and ^{17}O T_1 times, measured in 99% deuterium-labeled water with the natural abundance of ^{17}O nuclei, ranged from 0.2 to 0.57 s (275–310 K) for deuterium and from 2.5 to 17.2 ms (275–355 K) for oxygen. Nuclear quadrupole coupling constants at ^{17}O and ^2H in isolated water molecules are well known: $QCC_D = 308\,kHz$ ($\eta_D = 0.14$) and $QCC_O = 10.2\,MHz$ and ($\eta_O = 0.75$). However, owing to hydrogen bonding, the electronic structure of molecules and hence the quadrupole parameters at deuterium and oxygen change. Under these conditions, determination of the rotational correlation times from the ^2H, ^{17}O T_1 measurements requires accurate knowledge of the temperature-dependent QCC values. The B3LYP/6-31+G* calculations of oxygen and proton chemical shifts and also oxygen and deuterium quadrupole coupling constants in nine H-bonded water clusters, modeling the liquid water, have revealed the QCC/δ correlations:

$$QCC_D = -15.97\,\delta_H + 309.88 \tag{11.10}$$

$$QCC_O = 0.0893\,\delta_O + 9.9573$$

The temperature-dependent ^{17}O and ^2H chemical shifts, obtained experimentally, can be combined with Equations (11.10) to give the QCC values as a function of temperature:

$$QCC_D = 0.134\,T + 206.4 \tag{11.11}$$

$$QCC_O = 0.00405\,T + 7.74$$

As mentioned above, the ^2H and ^{17}O T_1 times have been measured between 275 and 350 K. According to Equations (11.11), in this temperature range, the oxygen and deuterium quadrupole coupling constants vary from 8.85 to 9.16 MHz and from 243 to 253 kHz, respectively. Then, these QCC values and the corresponding T_1 times give the τ_C values: $\tau_C(^2H_2O)$, ranging from 5.8 ps (275 K) to 0.86 ps (350 K), and $\tau_C(D_2\,^{17}O)$, changing from 4.4 ps (275 K) to 0.64 ps (350 K). As can be seen, the $\tau_C(^2H_2O)$ and $\tau_C(D_2\,^{17}O)$ times are not identical, and thus molecular motions in liquid water are anisotropic.

11.5 Two-dimensional T_1 and $T_{1\rho}$ NMR Experiments

Two-dimensional (2D) NMR, providing magnetization transfers and observation of chemical shift correlations, has great advantages in studies of complex molecular systems, such as polymers or biologically important molecules. Motion of these systems is an important aspect of their chemical behavior. It follows from Figure 4.2 that relatively slow molecular motions can be probed by T_1 and $T_{1\rho}$ measurements. In the case of complex systems, the T_1 or $T_{1\rho}$ pulse sections can be incorporated into the basic 2D NMR pulse sequences [10]. Under these conditions, volumes of cross-peaks, observed in the chemical shift correlation NMR spectra, will depend on time delays τ varied in the T_1 or $T_{1\rho}$ sections. Then, fitting the volumes of these cross-peaks, for example, to monoexponential decays (i.e. $V(\tau) = V_{(\tau=0)} \exp(-\tau/T)$, where T is the relaxation time) gives the T_1 or $T_{1\rho}$ values [10]. Note that these experiments require application of a three-channel NMR probes.

Figure 11.7 shows the 2D NMR pulse sequences employed for T_1 or $T_{1\rho}$ measurements at the deuterium frequency in aqueous solutions of the $^{13}C/^2H$ double-labeled 2'-deoxyadenosine, 1A, and the corresponding thymidine derivative, 2T (Figure 11.8). Here pulse sections A and B serve for T_1 and $T_{1\rho}$ determinations. Since the compounds contain ^{13}CHD groups, the

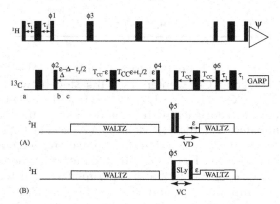

Figure 11.7 Pulse sequence applied for the measurements of T_1 (section A) and $T_{1\rho}$ (section B) relaxation times for ^2H nuclei using a TXO probe. Narrow and wide pulses correspond to 90° and 180°, respectively. The τ_1 time was 1.79 ms. The Δ time was set to 3.57 ms. The ε value was equal to 11.4 ms The T_{CC} delay has been inserted to eliminate $^{13}C-^{13}C$ coupling. (Reproduced with permission from T. V. Maltseva, A. Foldesi, J. Chattopadhyaya. *Magnetic Resonance in Chemistry* 1999; **37**: 203. © John Wiley & Sons, Ltd.)

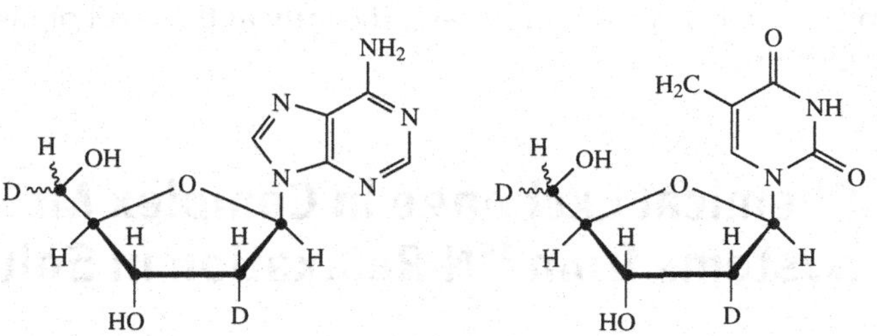

Figure 11.8 Selectively deuterated and ^{13}C enriched (solid black circles) 2'(R/S), 5'(R/S)-$^2H_2$-1',2',4',5'-$^{13}C_5$-2'-deoxyadenosine (**1A**) and its thymidine derivative (**2T**). (Reproduced with permission from T. V. Maltseva, A. Foldesi, J. Chattopadhyaya, *Magnetic Resonance in Chemistry* 1999; **37**, 203. © John Wiley & Sons, Ltd.)

Table 11.7 Comparison of T_1 and $T_{1\rho}$ times for ^2H nuclei in $^2H(C\text{-}2')$ groups of compounds 1A, 2T (Figure 11.8) and oligo-DNA in aqueous solutions at 11.7 T and 298 K

Compound	T_1 (s)	$T_{1\rho}$ (s)
1A	39.7 ± 0.6	39.7 ± 0.6
2T	57.2 ± 0.5	55.0 ± 0.9
Oligo-DNA		
^3A	6.34 ± 0.20	1.69 ± 0.14
^4T	6.72 ± 0.35	1.62 ± 0.09
^7A	7.39 ± 0.37	1.36 ± 0.07
^8T	9.22 ± 0.44	1.83 ± 0.099

NMR experiments have been performed with the $^1H-^{13}C-^2H-^{13}C-^1H$ polarization transfer at C-2' centers (see the step between points a and b where $\tau_1 = 1/4J(^{13}C-^1H)$, $\Delta = 1/2J(^{13}C-^1H)$ and $\varepsilon = 1/4J(^{13}C-^2H)$). The compounds have been enriched with ^{13}C and therefore the $^{13}C-^{13}C$ splitting should be eliminated (see the T_{CC} delay). WALZ and GARP pulses have been applied for decoupling of ^2H and ^{13}C nuclei, respectively. Finally, the volumes of the cross-peaks have been fitted to monoexponential decays. The ^2H T_1 and $T_{1\rho}$ data, collected by these experiments, are listed in Table 11.7. As can be seen, compounds 1A and 2T show very similar T_1 and $T_{1\rho}$ times. This result is expected because both molecules are not big, and undergo fast molecular motions on the frequency scale of ^2H NMR. The situation changes when compounds 1A and 2T are incorporated into the oligo-DNA: $5'd(^1C^2G^3A^4T^5T^6A^7A^8T^9C^{10}G)_2{}^{3'}$ (3). As seen, the ^2H $T_{1\rho}$ times in the nucleotide residues of the oligo-DNA are much shorter than ^2H T_1. This effect reveals the presence of slow motions in the big molecules. Add that the $T_1/T_{1\rho}$ ratios in all the nucleotide residues are different. Thus, the dynamic behavior of A and T residues differs significantly at the level of fast and slow molecular motions.

11.6   Chemical Exchange in Complex Molecular Systems from ^{15}N Relaxation in Solution

We have already noted that slow chemical exchanges lead to broadened NMR signals and the linewidths do not depend on the magnetic field strength. In contrast, the broadenings become field-dependent at fast exchanges, resulting in averaged signals. If molecules have relatively small molecular weights, and populations of the exchanging states are comparable, then the presence of chemical exchange can be easily established by appropriate methods (see

Chapter 3). In the case of systems of high molecular weights (polymers, biomolecules etc.), this task is not simple. In fact, owing to the presence of backbone and side-chain motions, multiple conformation states or numerous distinct magnetic environments, such systems give very complicated NMR spectra. All these circumstances require other approaches to identification of the exchanges.

Dynamics of protein molecules in solutions covers a large region of characteristic times lying between 10^{-12} and 10^4 s. Fast molecular motions with the correlation times from 10^{-12} to 10^{-9} s are in focus of T_1, T_2 and NOE measurements. Slow motions with τ_C values of 10^{-6}–10^{-3} s can be studied by $T_{1\rho}$ experiments. Rates of chemical exchanges in proteins are comparable with frequencies of molecular motions and their correlation times are between microseconds and milliseconds. Since the chemical exchanges are critical for protein biological function, it is important to reveal molecular sites, involved into the exchanges. In spite of the fact that protein molecules can be enriched with ^{15}N or ^{13}C, helping the detection of NMR spectra, resonance lines of exchanging molecular sites are strongly broadened and their intensities are weakened.

Usually, the contribution of a chemical exchange R_{EX} to the transverse relaxation rate $R_2(^{15}N)$, is:

$$R_{EX} = R_2 - R_2{}^0 \tag{11.12}$$

where $R_2{}^0$ is the natural transverse relaxation rate, governed by dipole–dipole and/or chemical shift anisotropy interactions. In the case of protein molecules, populations of exchanging states are different and therefore, even at a slow chemical exchange, only one resonance (corresponding to a major population) can be detected. Under these conditions, it is important to establish whether the R_2 is greater than the $R_2{}^0$ expected for relaxation mechanisms other than chemical exchange [11].

Details of pulse sequences, applied for studies of chemical exchanges in solutions of proteins, can be found in the literature [11]. With these pulse sequences the relaxation rates R_2 (rather than lineshapes) are measured via line intensities in NMR decays where evolutions by chemical shift and scalar coupling are suppressed. The latter is reached by incorporation of the Hahn-spin-echo, Carr–Purcell–Meiboom–Gill and spin-locking ($R_{1\rho}$) pulse sections. However, it must be emphasized that the train of 180° pulses, operating in CPMG experiments (see Equation 2.12), partially reduces the chemical exchange contributions. Therefore, measurements of R_2 values, as phenomenological relaxation rates, by the Hahn-echo sequence:

$$90°x'-\tau-180°y'-\tau \text{ (echo detection)}- \tag{11.13}$$

are preferable. The R_{EX} rates are calculated via Equation (11.12), when the $R_2{}^0$ and R_2 values are determined. In turn, the $R_2{}^0$ values can be obtained by experiments with full suppression of exchange effects. If, for some reason, an exchange is much slower than 1 ms, application of the CPMG pulses with delay times τ of 1 ms completely suppresses exchange effects. Alternatively, the $R_2{}^0$ values can be obtained on the basis of magnitudes $(R_2 - R_1/2)$, measured by the variable-field experiments. Here R_1 is the longitudinal relaxation rate and the R_2 rate is determined with help of the CPMG pulses [12]. This approach, however, requires application of three or more static magnetic fields. Finally, when exchange processes operate on the microsecond time scale, the $R_2{}^0$ rates can be obtained as $1/T_{1\rho}$ values, measured by spin-locking experiments.

It follows from the above considerations that the choice of pulse sequences applied for $R_2{}^0$ measurements depends on the rates of exchange. In the absence of such information, $R_2{}^0$ values can be obtained via cross-correlations between the ^{15}N CSA and $^{15}N–^1H$ dipole–dipole relaxation mechanisms. Phenomenologically, these cross-correlations are observed as different transverse relaxation rates measured for two lines in the $^{15}N–H$ doublets. The 2D NMR pulse sequence, applied in these experiments, includes the Hahn-echo pulse section, leading to determination of the transverse cross-correlation rate, σ_2 (DD,CSA) [11]. Technically the pulse sequence operates twice for each Hahn-echo time. The first experiment selects for in-phase coherence with signal intensity I_1. The second experiments selects for anti-phase coherence with signal intensity I_2. Then, the σ_2 (DD,CSA) value is calculated via the equation:

$$\sigma_2(\text{DD,CSA}) = -(1/\tau_{HE})\tanh^{-1}[I_1(\tau_{HE})/I_2(\tau_{HE})] \qquad (11.14)$$

where τ_{HE} is the Hahn-echo time equal to $n/J(^{15}N–^1H)$ and n is a positive integer.

11.7 R_1/R_2 Method

In spite of the essentially different origin of the spin–spin and spin–lattice relaxation mechanisms, fast molecular motions ($\omega^2\tau_C{}^2 \ll 1$) equalize T_1 and T_2 times. However, the equality $T_1 = T_2$, is violated at slower molecular motions with τ_C values ranging typically between nanoseconds and milliseconds. Under these conditions, the relaxation rates R_1 and R_2, measured at different temperatures, can be computed to give all the unknown parameters in the relaxation equations [13]. For example, relaxation rates of quadrupolar nuclei are:

$$R_1(Q) = 1/T_1\,(Q) = (3/100)\pi^2(2I + 3)[I^2(2I - 1)]^{-1}(e^2q_{zz}Q/h)^2(1 + \eta^2/3)$$
$$\times\ [2\tau_c/(1 + \omega_Q{}^2\tau_c{}^2) + 8\tau_c/(1 + 4\omega_Q{}^2\tau_c{}^2)] \qquad (11.15)$$

and:

$$R_2(Q) = 1/T_2(Q) = (3/200)\pi^2(2I + 3)[I^2(2I - 1)]^{-1}(e^2q_{zz}Q/h)^2(1 + \eta^2/3)$$
$$\times\ [6\tau_c + 10\tau_c/(1 + \omega_Q{}^2\tau_c{}^2) + 4\tau_c/(1 + 4\omega_Q{}^2\tau_c{}^2)] \qquad (11.16)$$

It is obvious that, at $\omega\tau_C > 1$, T_2 is shorter than T_1. Then, a combination of Equations (11.15) and (11.16) leads to the expression:

$$R_1(Q)/2R_2(Q) = [1/(1 + \omega_Q^2\tau_c^2) + 4/(1 + 4\omega_Q^2\tau_c^2)]$$
$$/[3 + 5/(1 + \omega_Q^2\tau_c^2) + 2/(1 + 4\omega_Q^2\tau_c^2)] \quad (11.17)$$

In turn, the expression can be rewritten:

$$\omega_Q^4\tau_c^4 + [3.0833 - 0.6667(2R_2(Q)/R_1(Q))]\omega_Q^2\tau_c^2$$
$$- 0.4167(2R_2(Q)/R_1(Q)) + 0.8333 = 0 \quad (11.18)$$

enabling calculation of the correlation time. Also, at $R_2 > R_1$, the solution of Equation (11.18) is a single positive τ_C value. Then, assuming $\eta = 0$, this value can be used for calculations of the $e^2 q_{zz}Q/h$ constants via Equation (11.15). It should be noted that the condition $\omega\tau_C > 1$ can be realized, for example, in viscous media.

The R_1/R_2 approach has been used to analyze ^{27}Al relaxation in a molten salt formed by LiCl and ethyl aluminum dichloride in a 1:1 ratio. Figure 11.9

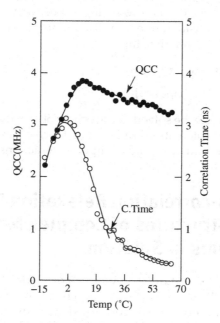

Figure 11.9 ^{27}Al nuclear quadrupole coupling constants (QCC, full circles) in MHz and effective molecular motion correlation times (open circles) for 1:2 LiCl – EtAlCl$_2$ versus temperature (°C). (Reproduced with permission from W. R. Carper. *Concepts in Magnetic Resonance* 1999; **11**: 51. © John Wiley & Sons, Ltd.)

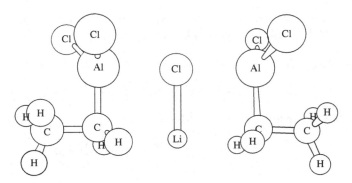

Figure 11.10 The MOPAC (PM3)-optimized structure of the complex, formed at an addition of LiCl to EtAlCl$_2$. (Reproduced with permission from W. R. Carper, *Concept in Magnetic Resonance* 1999; **11**, 51. © John Wiley & Sons, Ltd.)

shows the computed aluminum quadrupole coupling constants in the temperature region between $-12°$ and $65°$C. As can be seen, on cooling (from $65°$ to $10°$C), the $e^2q_{zz}Q/h$ value slightly increases. Then, decreasing the temperature from $10°$ to $-12°$C leads to the opposite effect. The data allow one to conclude that the insertion of LiCl into an EtAlCl$_2$ dimer (Figure 11.10) alters the C$_{2h}$ symmetry of the dimer, producing a species of a higher symmetry.

By analogy, these considerations, applied for proton–proton dipole–dipole interactions, lead to the relationship:

$$R_1/R_2 = [2/(1 + \omega_H^2\tau_c^2) + 8/(1 + 4\omega_H^2\tau_c^2)]/[3 + 5/(1 + \omega_H^2\tau_c^2)$$
$$+ 2/(1 + 4\omega_H^2\tau_c^2)] \tag{11.19}$$

Again, this equation can be solved iteratively to obtain the τ_c value. The latter eventually provides calculations of H–H distances. Similar expressions can be deduced for heteronuclear dipole–dipole interactions [13].

11.8 Cross-correlation Relaxation Rates and Structures of Complex Molecular Systems in Solution

In section 11. 6 we have shown how cross-correlation relaxation rates can be used to establish and investigate the chemical exchanges in complex molecular systems. Here we are concerned with the structural aspects of cross-relaxation measurements. Knowledge of backbone dihedral angles ϕ in protein molecules allows one to deduce their secondary structure in solution. In turn, the dihedral angles are related to angles θ, characterizing mutual orientations of two successive C–H bond vectors (see Figure 11.11).

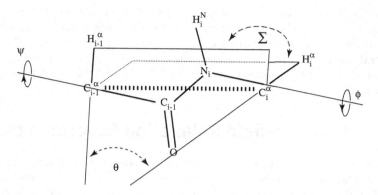

Figure 11.11 A fragment of the backbone of a protein. The θ angle is subtended between two bond vectors $^{13}C^{\alpha}(i\text{-}1)$–$H^{\alpha}(i\text{-}1)$ and $^{13}C^{\alpha}(i)$–$H^{\alpha}(i)$ in successive amino acids. The Σ angle is the dihedral angle between the planes defined by the atoms $H^{\alpha}(i\text{-}1)$, $^{13}C^{\alpha}(i\text{-}1)$, $^{13}C^{\alpha}(i)$ and $^{13}C^{\alpha}(i\text{-}1)$, $^{13}C^{\alpha}(i)$, $H^{\alpha}(i)$. (Reproduced with permission from E. Chiarparin, P. Pelupessy, R. Ghose, G. Bodenhausen. *Journal of the American Chemical Society* 2000; **122**: 1758. © 2000 American Chemical Society)

When proteins are labeled with ^{15}N and ^{13}C isotopes, the θ angles dictate the cross-correlation relaxation rates, $R_1(\theta)$, originating from interference between carbon–proton dipolar couplings in two neighboring C–H bonds according to:

$$1/T_1(\theta) = R_1(\theta) = (\mu_0\hbar/4\pi)^2 r(\text{H–C})^{-6}\gamma_H^2\gamma_C^2(2/5)$$
$$\times\, S^2\tau_C(1/2)(3\cos^2\theta - 1) \tag{11.20}$$

where S^2 is the Lipari–Szabo order parameter and $r(\text{H–C})$ can be taken as 1.12 Å [14].

Details of 2D NMR pulse sequences, serving for $R_1(\theta)$ determinations, can be found in the literature [14]. Here, it is be noted that they include magnetization transfers from the amid protons H(N) to ^{15}N and then to $^{13}C(\alpha)$ nuclei in the same and the previous residue. Then, the $R_1(\theta)$ rate can be obtained via the equation:

$$I(2)/I(1) = [\exp(R_1(\theta)T) - \exp(-R_1(\theta)T)]/$$
$$[\exp(R_1(\theta)T) + \exp(-R_1(\theta)T)] = \tanh(R_1(\theta)T) \tag{11.21}$$

where $I(1)$ and $I(2)$ are the signal intensities measured by two independent 2D NMR experiments, and T is the relaxation delay. In these experiments, the evolution under scalar couplings during the relaxation time period is different. In experiment (1), all scalar couplings are refocused. In experiment (2), two $J(\text{C–H})$ couplings are active. According to Chiarparin *et al.* [14], the relaxation delays T should be either very short (~6 ms) or a multiple of the $1/J(C\alpha\text{-}C\beta)$ values ($\sim$25 ms) in order to avoid losses because of evolution by the $^{13}C\text{-}^{13}C$ coupling. The measurements and an analysis of the $R_1(\theta)$ values in each pair of the $C\text{-}H$ vectors result in calculations the dihedral angles. Finally, by analogy, the dihedral angles can be obtained by studies of cross-correlations between the $^{15}N\text{-}H$ and $^{13}C\text{-}H$ dipole–dipole interactions.

11.9 Variable-field Relaxation Experiments

In contrast to small molecules, systems of high molecular weight can violate the condition of fast molecular tumbling, $\omega^2\tau_c^2 \ll 1$, even at high tempera-ture. For this reason, nuclear relaxation is field dependent and therefore the variable-field relaxation measurements become a useful tool for studies of molecular dynamics in solution. Figure 11.12 shows the variable-field relax-ation rates R_1, calculated by Redfield for two favorite nuclei in biopolymers, ^{15}N and ^{31}P, according to the dipolar and CSA mechanisms [15]. The ^{15}N R_1 curves (Figure 11.12a) were obtained for a typical amide fragment of a theoretically rigid protein with a slowly exchanging proton and a rota-tional correlation time of 5 ns. The ^{31}P R_1 curves (Figure 11.12b) reflect the relaxation behavior of a phosphate group in a small DNA duplex with rotational correlation time of 4.2 ns. The correlation time of an internal fast

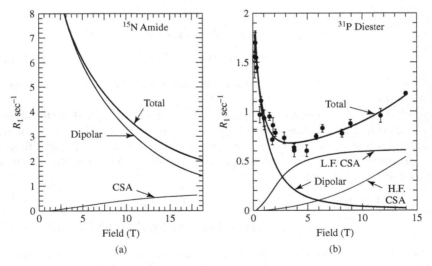

Figure 11.12 Field dependence of NMR relaxation rates expected for macromolecules: (a) ^{15}N relaxation rate for a typical peptide amide in a protein; (b) ^{31}P relaxation rates for a phosphodiester group in small DNA duplex. (Reproduced with permission from A. G. Redfield. *Magnetic Resonance in Chemistry* 2003; **41**: 753. © John Wiley & Sons, Ltd.)

motion in the duplex was taken as 200 ps. The CSA contributions have been calculated in terms of the $\Delta\sigma(^{15}N)$ and $\Delta\sigma(^{31}P)$ values equal to 160 ppm. The curve H.F.CSA corresponds to the ^{31}P CSA relaxation rate for a fast internal motion of the phosphate group and the curve L.F.CSA is the ^{31}P CSA contribution from the remaining CSA after averaging over this motion due to overall molecular tumbling. According to the calculations, the two nuclei show different variable-field phenomena and thus they can potentially provide important information about molecular mobility. For example, the influence of fast internal motions on ^{31}P relaxation rate is particularly strong at high fields. It is obvious that such effects could be studied with the use of several NMR spectrometers working at different NMR frequencies. However, a detailed analysis of the spectrum of motions is impossible in the absence of measurements at low magnetic fields. This statement is well illustrated by experimental measurements carried out for the low-field signal in the ^{31}P NMR spectra of an aqueous solution of the DNA octamer duplex (dGpGpApApTpTpCpC)$_2$ (see the points in Figure 11.12b). As can be seen, these points are well fitted to the motional model of a phosphate group suggested above. Finally, note that technically the measurements can be made with field-cycling NMR spectrometers containing switched coil magnets or an apparatus providing a pneumatic shift of a sample from the center of a standard magnet to any position, under computer control [15].

Besides information about the character of molecular motions, the variable-field experiments can be used in studies of chemical exchange. As has been mentioned, the $T_{1\rho}$ times are sensitive to exchange processes occurring in solutions of proteins on the time scale between μs and ms. Note that rotating-frame experiments lock magnetization of a sample along the direction of the effective radiofrequency field operating at frequency ω_E. Under these conditions $R_{1\rho}$ values depend on the amplitude of the applied field. Then, measuring the $R_{1\rho}$ as a function of ω_E (so-called relaxation dispersion), leads to determinations of the kinetic parameters of exchange. In fact, the contribution R_{EX} to the total relaxation rate (see Equation 11–12), caused by the presence of a two-site exchange, is:

$$R_{EX} = k_{EX}P_AP_B \, \delta_{AB}{}^2/(k_{EX}{}^2 + \omega_E{}^2) \qquad (11.22)$$

where k_{EX} is the exchange constant, P_A and P_B are populations of the two states and δ_{AB} is the chemical shift difference. Note that the variable-frequency $R_{1\rho}$ experiments can be made on ^{13}C or ^{15}N nuclei of protein molecules, with variation in the ω_E between 150 and 1000 Hz and suppression of ^1H coupling and cross-relaxation processes [16].

Finally Figure 11.13 shows an example of proton variable-frequency experiments. Here the frequency-dependent ^1H R_1 relaxation rates are measured for the protons of water and DMSO in a serum albumin protein solution [17]. The relaxation dispersion profiles reflect the situation when few solvent molecules are binding to protein molecules and this effect is observed by

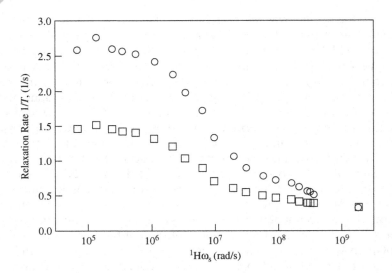

Figure 11.13 Nuclear magnetic relaxation dispersion profiles for water protons and protons of DMSO recorded simultaneously in an 11.7% solution of bovine serum albumin, containing 11.7% DMSO and 8.5% H_2O in D_2O at room temperature. The upper curve corresponds to the water protons; the lower curve corresponds to the CH_3 protons. (Reproduced with permission from S. Wagner *et al. Journal of Magnetic Resonance* 1999; **140**: 172.)

relaxation of protons in bulk water (or DMSO) due to fast exchanges between the free and bonded states. Methods of treatment of the relaxation dispersion data, providing determinations of the number of water molecules binding to proteins can be found in the literature [18].

Bibliography

1. S. M. Socol, D. V. Meek. *Inorgica Chimica Acta* 1985; **101**: L45.

2. A. Bodor, I. Banyai, J. Kowalewski, J. Glaser. *Magnetic Resonance in Chemistry* 2002; **40**: 716; S. Molchanov, A. Gryff-Keller. *Magnetic Resonance in Chemistry* 2003; **41**: 788.

3. C. Sizun, P. Kempgens, J. Raya, K. Elbayed, P. Granger, J. Rose. *Journal of Organometallic Chemistry* 2000; **604**: 27.

4. W. R. Carper, J. L. Pflug, J. S. Wilkes. *Inorgica Chimica Acta* 1992; **193**: 201.

5. A. Bagno, S. Gerard, J. Kevelam, E. Menna, G. Scorrano. *Chemistry. A European Journal* 2000; **6**: 2915.

6. Y. Qu, N. Farrel. *Inorgica Chimica Acta* 1996; **245**: 265.

7. F. C. Marincola, M. Casu, G. Saba, A. Lai. *Chem Phys Chem. A European Journal of Chemical Physics, and Physical Chemistry* 2001; **2**: 569.

8. M. Casu, G. Saba, A. Lai, M. Luhmer, A. K. Mesmaeker, C. Moucheron, J. Reisse. *Biophysical Chemistry* 1996; **59**: 133.

9. J. Ropp, C. Lawrence, T. C. Farrar, J. L. Skinner. *Journal of the American Chemical Society* 2001; **123**: 8047.

10. T. V. Maltseva, A. Foldesi, J. Chattopadhyaya. *Magnetic Resonance in Chemistry* 1999; **37**: 203.

11. C. Wang, A. G. Palmer. *Magnetic Resonance in Chemistry* 2003; **41**: 866.

12. I. Q. H. Phan, J. Boyd, I. D. Cambell. *Biomolecular NMR* 1996; **8**: 369.

13. W. R. Carper, *Concepts in Magnetic Resonance*, 1999; **11**: 51.

14. E. Chiarparin, P. Pelupessy, R. Ghose, G. Bodenhausen. *Journal of the American Chemical Society* 2000; **122**: 1758.

15. A. G. Redfield. *Magnetic Resonance in Chemistry* 2003; **41**: 753.

16. F. Massi, E. Johnson, C. Wang, M. Rance, A. G. Palmer III. *Journal of the American Chemical Society* 2004; **126**: 2247.

17. S. Wagner, T. R. J. Dinesen, T. Rayner, R. G. Bryant. *Journal of Magnetic Resonance* 1999; **140**: 172.

18. A. Van-Quynh, S. Willson, R. G. Bryant. *Biophysical Journal*, 2003; **84**: 558.

12 Paramagnetic NMR Relaxation

Organic and inorganic species, containing paramagnetic metal ions, are widely used in NMR as shift reagents. Owing to action of unpaired electrons, the chemical shift differences of nuclei increase and, for this reason, the complicated NMR spectra are simplified. These species play an important role in determinations of molecular sites interacting with metal centers or investigations of different covalent and noncovalent intermolecular interactions. Qualitative and quantitative studies of exchange processes are the third field for applications of paramagnetic complexes. Such investigations are carried out over a wide range of compounds, from small organic molecules to metalloproteins. In addition, paramagnetic ions are cofactors for a number of biologically important processes and therefore structures of paramagnetic complexes are of great interest. Finally, systems, containing paramagnetic metal ions, are an important class of contrast agents for magnetic resonance imaging, currently applied for clinical diagnosis.

12.1 Theoretical Basis of Paramagnetic Relaxation Enhancement

The rate of nuclear relaxation increases in the presence of a dissolved paramagnetic ion and this effect is named the NMR paramagnetic relaxation

Practical NMR Relaxation for Chemists Vladimir I. Bakhmutov
© 2004 John Wiley & Sons, Ltd   ISBNs: 0-470-09445-1 (HB); 0-470-09446-X (PB)

enhancement. The physical origin of the phenomenon consists of interaction between magnetic moments of an electron and a nucleus:

$$\mu_S = -g_e \, \beta_e \, S \tag{12.1}$$

$$\mu_I = \gamma \, \hbar \, I \tag{12.2}$$

where g_e is the electron g–factor (2.00232) and β_e is the Bohr magneton ($9.2741 \times 10^{-24} \, J \, T^{-1}$). In the presence of the external magnetic field, both magnetic moments precess around its direction and thermal fluctuations in the dipole–dipole interactions cause nuclear relaxation. Such a situation occurs, for example, in organic Cu(II) complexes where an unpaired electron is located in the Cu atom ($S = 1/2, I = 3/2$) and target nuclei belong to a metal-bound ligand.

Theoretically, for the paramagnetic Cu(II) complexes, rates of the nuclear relaxation, $R_{1,2}(I)$, are expressed by:

$$1/T_1(I) = R_1(I) = R_1(I)^{\mathrm{DIP}} + R_1(I)^{\mathrm{CON}} + R_1(I)^{\mathrm{CURIE}}$$
$$1/T_2(I) = R_2(I) = R_2(I)^{\mathrm{DIP}} + R_2(I)^{\mathrm{CON}} + R_2(I)^{\mathrm{CURIE}} \tag{12.3}$$

where R^{DIP}, R^{CON} and R^{CURIE} are the dipolar, contact and Curie contributions, respectively [1]. The dipolar contribution is given by the Solomon equations:

$$R_1(I)^{\mathrm{DIP}} = (2/15) \, (\mu_0/4\pi)^2 \, \gamma_I^2 \, g_e^2 \, \beta_e^2 \, S(S+1) \, r(I{-}S)^{-6}$$
$$\times \, [3\tau_C/(1 + \omega_I^2\tau_C^2) + 7\tau_C/(1 + \omega_E^2\tau_C^2)]$$
$$R_2(I)^{\mathrm{DIP}} = (1/15) \, (\mu_0/4\pi)^2 \, \gamma_I^2 \, g_e^2 \, \beta_e^2 \, S(S+1) \, r(I{-}S)^{-6}$$
$$\times \, [4\tau_C + (3\tau_C/(1 + \omega_I^2\tau_C^2) + 13\tau_C/(1 + \omega_E^2\tau_C^2)] \tag{12.4}$$

where S is the electron spin, ω_I and ω_E are the nuclear and electron Larmor frequencies, $r(I{-}S)$ is the electron-nucleus distance and τ_C is the correlation time. The correlation time can be defined as $1/\tau_C = 1/\tau_R + 1/\tau_E$, where τ_E is the electron spin relaxation time and τ_R is the molecular rotational correlation time. It is obvious that at $\tau_R \gg \tau_E$, the τ_C is controlled by the τ_E and vice versa, the τ_C is governed by the τ_R at $\tau_E \gg \tau_R$. Table 12.1 lists the electron relaxation times for some paramagnetic metal ions [2] where τ_{1E} times cover the range from 10^{-7} to 10^{-13} s. Note that the molecular rotational correlation times, being a function of the size of molecules, are between 10^{-11} and 10^{-6} s.

The contact contributions, arising from delocalization of the unpaired spin density in a metal-bound ligand, are given by the expressions:

$$R_1(I)^{\mathrm{CON}} = (2/3) \, S(S+1) \, (A/\hbar)^2 \, (\tau_{E2}/(1 + \omega_E^2\tau_{E2}^2)$$
$$R_2(I)^{\mathrm{CON}} = (1/3) \, S(S+1) \, (A/\hbar)^2 \, (\tau_{E1} + (\tau_{E2}/(1 + \omega_E^2\tau_{E2}^2)) \tag{12.5}$$

where A is the hyperfine coupling constant, characterizing the scalar electron–nucleus interactions, and τ_{E1} and τ_{E2} are the longitudinal and transverse

Table 12.1 Electron relaxation times for different paramagnetic metal ions, τ_E, and linewidths $\Delta\nu$ expected for stable paramagnetic complexes

Ion	Electronic configuration	S	τ_E (s)	$\Delta\nu$ (Hz)
Ti^{3+}	d^1	$1/2$	$10^{-10}-10^{-11}$	20–200
VO^{2+}	d^1	$1/2$	10^{-8}	10 000
V^{3+}	d^2	1	10^{-11}	50
Mn^{3+}	d^4	2	$10^{-10}-10^{-11}$	150–1500
Ni^{2+}	d^8 5–6 coord	1	10^{-10}	500
	d^8 4 coord	1	10^{-12}	5
Cu^{2+}	d^9	$1/2$	$10^{-8}-10^{-9}$	1000–5000
Gd^{3+}	f^7	$7/2$	$10^{-8}-10^{-9}$	20 000–200 000

electron spin relaxation times, respectively. Emphasize that by the definition, reorientational molecular motions do not affect the contact terms. Finally at $S = 1/2$, the Curie contribution is:

$$R_1(I)^{CURIE} = (1/40)\,(\mu_0/4\pi)^2\,(\omega_I^2\,g_e^4\beta_e^4/k^2T^2)\,r(I-S)^{-6}$$
$$\times\,(3\tau_R/(1+\omega_I^2\tau_R^2))$$
$$R_2(I)^{CURIE} = (1/80)\,(\mu_0/4\pi)^2\,[\omega_I^2\,g_e^4\,\beta_e^4/k^2T^2)\,r(I-S)^{-6}$$
$$\times\,(4\tau_R + 3\tau_R/(1+\omega_I^2\tau_R^2))] \tag{12.6}$$

where k is the Boltmann constant and T is the absolute temperature.

As it follows from the theory, structural interpretations of paramagnetic relaxation rates require previous separations of three contributions $R_1(I)^{DIP}$, $R_1(I)^{CURIE}$ and $R_1(I)^{CON}$ because only two of them contain the $r(I-S)$ parameter. Comparison of Equations (12.4) and (12.6) shows that the Curie contribution to the $R_1(I)$ relaxation rate can be significant at $\tau_C = \tau_E$ and $\tau_E \ll \tau_R$. If the τ_C time (Equation 12.4) is controlled by τ_R, then the Curie contribution is negligibly small. Actually, it is easy to show that under these conditions $R_1(I)^{DIP}/R_1(I)^{CURIE} = \sim1000$. In practice this situation occurs in the case of small and medium-size ligands ($\tau_R = 10^{-11}-10^{-10}$ s) coordinated with the Cu^{2+} or VO^{2+} ions, the τ_E values of which are of the order of $10^{-8}-10^{-9}$ s (see Table 12.1). On the other hand, such a favorable τ_R/τ_E ratio can result in a 'negative spectroscopic effect.' In fact, if the rotational mechanism dominates the correlation time τ_C, i.e. $\tau_C = \tau_R$, and the latter is quite long, then NMR lines of target nuclei broaden considerably. For example, compare in Table 12.1 the broadening effects, which can be caused by the coordination with Ti^{3+} and VO^{2+} ions.

The contact contribution in Equation (12.3) can be evaluated experimentally by comparison of the $R_1(I)$ and $R_2(I)$ times measured in the intermediate motional region, i.e. at $\omega_E\tau_C > 1 > \omega_I\tau_C$. If the contact mechanism actually dominates, then the $R_2(I)$ rate is much higher than $R_1(I)$. In the slow- or fast-motion regimes, the $R_1(I)$ and $R_2(I)$ times are not informative in this context. In such cases, the upper limit of the contact contribution is usually estimated via Equation (12.5) for the known τ_E and A values. Then the obtained magnitude can be compared with the $R_1(I)$ rate, measured experimentally.

Solomon's theory adequately describes nuclear relaxation occurring in organic radicals where electrons, centered on paramagnetic metal ions, have spins of $1/2$ (such as Cu(II)). The main principle of this theory consists of a Zeeman quantization of both the nuclear and electron spin motions in the presence of the external magnetic field B_0:

$$H_I = H_{Zeem} = -\mu_I B_0$$

$$H_S = H_{Zeem} = -\mu_S B_0 \tag{12.7}$$

In other words, the spin vector S, like the nuclear vector I, undergoes a Larmor precession around the B_0 axis. However the assumption is physically inappropriate for electron spins $S \geqslant 1$ [3] taking place in many transition metal ions in their regular oxidation states (see, for example, ions Ni^{2+}, Mn^{3+} or Gd^{3+} in Table 12.1). Owing to orbital contributions, electrons with $S \geqslant 1$ are subject to the zero-field splitting interactions H_{ZFS}. When the zero-field splitting is comparable to or larger than the Zeeman splitting, the electron spin motion becomes complicated and quantized along the molecule-fixed coordinate axis, but not along the external magnetic field B_0. The latter has a great influence on nuclear relaxation. As a result, the zero-field splitting interactions and the Zeeman interactions are summarized:

$$H_S = H_{Zeem} + H_{ZFS} \tag{12.8}$$

Theoretically, in the zero-field splitting limit, i.e. at $H_{Zeem} \ll H_{ZFS}$, nuclear paramagnetic relaxation differs strongly from that in the Zeeman limit. For example, in the case of electrons with spins $S = 1$ and a cylindrical symmetry of the zero-field splitting tensor, the paramagnetic relaxation enhancement is a sum of longitudinal (R_{1Z}) and transverse ($R_{1\perp}$) contributions, i.e. $R_1 = R_{1Z} + R_{1\perp}$, where:

$$R_{1Z} = (8/3)\,(\mu_0/4\pi)^2\,g_e{}^2\beta_e{}^2\,\Phi(\theta)\,r(I\text{-}S)^{-6}\,J(\omega_I) \tag{12.9}$$

$$\Phi_Z(\theta) = (1/3)(1 + P_2\cos(\theta))$$

$$R_{1\perp} = (4/3)(\mu_0/4\pi)^2 g_e{}^2\beta_e{}^2\Phi(\theta)r(I\text{-}S)^{-6}J(\omega_D)$$

$$\Phi_\perp(\theta) = (1/6)(2 - P_2\cos\theta)$$

The θ in Equations (12.9) is the polar angle, formed by the interspin I–S vector and the principal axis of the zero-field splitting tensor (for the meaning of the

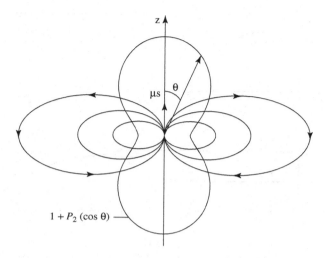

Figure 12.1 Angular function, $\Phi_Z(\theta) = (1/3)(1 + P_2 \cos(\theta))$, describing the angular variation of the square of the dipolar field of the dipole moment μ_S located at the origin and oriented along the Z-axis. (Reproduced from R. Sharp, L. Lohr, J. Miller. *Progress in Nuclear Magnetic Resonance Spectroscopy* 2001; **38**: 115, with permission from Elsevier)

$\Phi(\theta)$ function see Figure 12.1). Equations (12.9) emphasize the main feature of relaxation in the zero-field splitting limit. Actually, the relaxation rate in the Zeemans' limit depends on I–S distances, but not on the I–S vector orientation.

Since NMR relaxation is a function of the magnetic field strength, relaxation data can be presented as a field dispersion profile or, in other words, as a plot of R_1 versus the Zeeman field strength (or Larmor frequency) at a constant temperature. It has been mentioned already that in such experiments, NMR relaxation measurements are conducted over a wide range of static field strengths B_0. At the low end, the fields are several orders of magnitude lower than the typical magnetic fields applied for the high-resolution NMR experiments. Note that modern field-cycling magnets provide T_1 measurements in the B_0 range from 10^{-4} to 2 T. The upper end of this range can be extended up to $B_0 > 10$ T using the regular high-field spectrometers. Thus for electrons with $S \geqslant 1$, it is possible to create conditions where the zero-field splitting is much larger than the Zeeman interactions. Figure 12.2 illustrates the R_1 field-dispersion profiles, calculated by Sharp *et al.* [3] at different θ values for $S = 1$ and the cylindrical zero-field splitting tensor. As can be seen, the relaxation rates strongly depend on orientations of the interspin I–S vector at low magnetic fields. At the same time, the θ effects become invisible at high magnetic fields where the Zeeman interactions dominate. Hence, the magnetic field-dependent relaxation data can provide

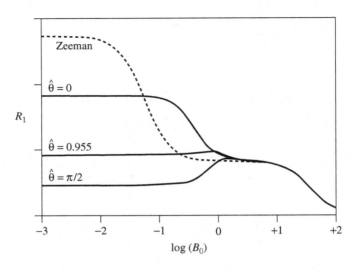

Figure 12.2 Relaxation rate R_1 dispersion profiles for $S = 1$ in the cylindrical zero-field splitting limit at three θ values. (Reproduced from R. Sharp, L. Lohr, J. Miller. *Progress in Nuclear Magnetic Resonance Spectroscopy* 2001; **38**: 115, with permission from Elsevier)

uniquely sensitive experimental probes of the finer aspects of the spin level structure. Finally, Sharp *et al.* emphasize that each spin value has unique aspects requiring separate consideration. These aspects can be found in the original paper [3].

Solomon's equations (12.4)–(12.6) show that the influence of unpaired electrons on nuclear relaxation depends of the nature of target nuclei. Since nuclei, other than protons, possess smaller magnetic moments, their dipolar interactions with unpaired electrons will be weaker. The rate of nuclear dipolar relaxation is proportional to the square of the γ_I constant and therefore these reducing effects are significant. For example, at identical nucleus–electron distances, the paramagnetic relaxation rate enhancements, observed for protons and ^{15}N nuclei, will be related as $\sim$100:1. Add also that line widths of ^{15}N resonances are expected to be significantly smaller. The same γ_I^2 effects on going from protons to nitrogen or oxygen nuclei are valid for contact contributions $R_{1,2}^{CONT}$ (see Equation 12.5). On the other hand, nitrogen and oxygen atoms can directly coordinate with paramagnetic metal centers, leading to increasing the electron spin density around ^{17}O, ^{15}N nuclei. Then the $A/\hbar$ value and the corresponding contact contribution increase with respect to protons in spite of the lower γ_I constants. Note that for such nuclei as ^{15}N, ^{19}F or ^{17}O, the $A/\hbar$ values are particularly big. In this case, the contact mechanism can become dominant and lead to greatly broadened NMR signals.

12.2 Paramagnetic Relaxation Rate Enhancements in the Presence of Chemical Exchange

Equations (12.4)–(12.6) indicate the influence of electron spin relaxation and molecular rotational reorientations on nuclear relaxation. The third type of motion affecting the relaxation is a chemical exchange with a correlation time τ_M. This exchange appears when the complex, formed by a paramagnetic ion and an organic ligand, is chemically unstable. Figure 12.3 compares ranges of typical values for τ_E, τ_R and τ_M correlation times. It is seen that the τ_M time, as a function of metal–ligand bond strength, can be indefinitely long (a slow chemical exchange) or as short as 10^{-10} s (a fast chemical exchange). It is obvious that a fast exchange maximally reduces the paramagnetic relaxation rate enhancements, measured experimentally.

In solutions, a nuclear spin, located in organic ligand L and interacting with a paramagnetic center, is involved in an exchange between the diamagnetic (free) and paramagnetic (metal- bound) states [1]:

$$L^{FREE} + M \Longleftrightarrow L^{M-BOUND} \qquad (12.10)$$

In the case of a 1:1 coordination, rate constants of the association and dissociation reactions τ_F^{-1} and τ_M^{-1} are mutually connected via:

$$\tau_F/\tau_M = P_F/P_M$$
$$P_F + P_M = 1$$

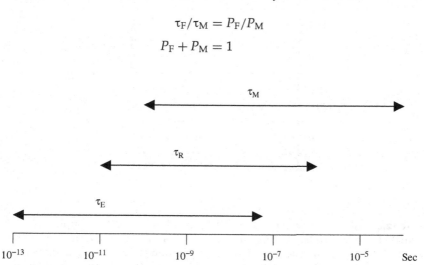

Figure 12.3 Electronic relaxation times τ_E in paramagnetic ions, molecular rotational correlation times τ_R of the paramagnetic complexes and correlation times τ_M of exchange between free and M-bound states

where P is the corresponding mole fraction. Finally, the temperature dependence of the τ_M is expressed by:

$$\tau_M^{-1} = (kT/h)\exp(-\Delta G^{\neq}/RT) \tag{12.11}$$

where $\Delta G^{\neq}$ is the free energy of activation.

It is obvious that exchange (12.10) averages the measured spin–lattice relaxation rates completely or partially. Phenomenologically, the relaxation rate enhancement $R_{1r} = 1/T_{1r}$, caused by a paramagnetic ion, is defined as:

$$R_{1r} = R_1 - R_{1F} \tag{12.12}$$

Here, the R_1 and R_{1F} values are measured in the presence and the absence of paramagnetic ions, respectively. Then, the rate of nuclear relaxation in the bound (paramagnetic) state can be determined via the R_{1r} by the equation:

$$R_{1r} = P_M R_{1M}[\tau_M^{-1}/(R_{1M} + \tau_M^{-1})] \tag{12.13}$$

Thus, the exchange rate is a limiting factor for a transfer of the paramagnetic influence to the diamagnetic state. An analysis of Equation (12.13) shows that only in the fast exchange limit, i.e. at $R_{1M} \ll \tau_M^{-1}$, the registered paramagnetic effect R_{1r} is directly related to R_{1M}. In other cases, interpretations of the R_{1r} values require independent knowledge of the exchange rates.

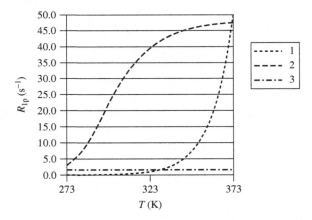

Figure 12.4 Temperature dependence of R_{1r} (see Equation 12.13 coupled 12.4, 12.11) calculated for a correlation time $\tau_C = 5 \times 10^{-10}$ s and a molar fraction of bound ligand of 0.01 by assuming $a = 0.82$ nm in Equation (1.14). Calculations performed at $R_{1M} \gg \tau_M^{-1}$ for $r(I-S) = 0.25$ nm and $\Delta G^{\neq}/R = 10\,000$ K (1), at $R_{1M} \sim \tau_M^{-1}$ for $r(I-S) = 0.35$ nm and $\Delta G^{\neq}/R = 8000$ K (2) and at $R_{1M} \ll \tau_M^{-1}$ for $r(I-S) = 0.60$ nm and $\Delta G^{\neq}/R = 6000$ K (3). (Reproduced with permission from E. Gaggelli, N. D'Amelio, D. Valensin, G. Valensin. *Magnetic Resonance in Chemistry* 2003; **41**: 877.)

One of the possible approaches to τ_M^{-1} determinations is based on variable-temperature R_{1r} measurements. In fact, under extreme narrowing conditions ($\omega_H \tau_R \ll 1$), the rotational correlation time τ_R depends on the temperature and the radius of molecules, as shown in Equation (1.14). The temperature dependence of the exchange rate τ_M^{-1} is governed by Equation (12.11). If both expressions are incorporated into Equation (12.13), then the plots of R_{1r} versus T will depend on the relaxation and exchange parameters simultaneously. Figure 12.4 illustrates the exchange effects on the R_{1r} curves obtained for parameters shown in the caption. Thus the R_{1r} temperature dependence is helpful in determining the exchange region and hence the exchange rate. However, this is impossible if the extreme narrowing conditions do not apply for the free ligand (i.e. $\omega_H \tau_R \geq 1$).

12.3 Structural Applications of Paramagnetic Relaxation Rate Enhancement

Since nuclear paramagnetic relaxation is very sensitive to the distances separating the target nuclei and paramagnetic centers, decreases in T_1 and T_2, measured for several nuclei in the same molecule, can range from negligible to very large. Thus, multinuclear relaxation is a powerful instrument for structural characterization of chemical compounds. However, measurements and interpretations of spin–lattice relaxation times are preferable because of intrinsic difficulties in measuring transverse relaxation times and the stronger influence of contact interactions on $R_2(I)$ values.

Protons, often the target nuclei in structural studies, can form multi-spin systems with a nonexponential NMR relaxation (see Chapter 3). In such cases, the T_1 times can be estimated by using selective or nonselective inversion recovery experiments [4]. In this connection, it is important to establish which T_1 values (selective or nonselective) are more reliable for $r(I-S)$ determinations. In fact, when all spins in a multi-spin system are simultaneously excited (nonselective excitation), they can give large cross-relaxation contributions. This effect is particularly significant in intermediate and slow-motion regimes. Banci and Luchinat [4] have analyzed the relaxation behavior of a spin system including two different protons A and B, and a paramagnetic metal center M. It has been established that at $R_1(A-M) = R_1(B-M)$, a *nonselective* inversion leads to an *exponential* recovery of the A (or B) magnetization. In addition, this recovery is absolutely identical with that obtained in the absence of the B proton i.e. for an isolated A/M pair. In contrast, a *selective* excitation of the A (or B) proton results in a *faster* and *nonexponential* magnetization recovery. It is probable that immediately after *nonselective inversion*, no magnetization transfer is possible between protons A and B and, hence, the recovery of the A magnetization is governed by interactions with the paramagnetic center M. The same calculations, carried out for $R_1(A-M) \neq R_1(B-M)$, lead to nonexponential magnetization recoveries after selective and nonselective inversion. However, *nonselective* recovery was a better approximation to the case of an isolated A/M pair. Thus, determinations of metal–nucleus distances by nonselective T_1 measurements are more reliable.

To illustrate the methodology of structural investigations, consider association of organic substrates with the paramagnetic Cu(II) complex 1, ({(R)-7,8,15,18-tetrahydro-5,10-bis(2′-methoxyethoxy-5′-methylphenyl)-7-(p-trime thylammoniophenyl)dibenzo[e,m]-[1,4,8,11]tetra-azacyclotetradecine-16,17-dionato(1-)-N⁶N⁹N¹⁵N¹⁸}copper(II) chloride [5]). Tables 12.2 and 12.3 show the $^1H T_1$ and $^1H T_2$ data, collected for aqueous solutions of proline (Figure 12.5) in the absence and in the presence of complex 1. As can be seen, the proline relaxation times are strongly affected by addition of complex 1. The effects are particularly significant for the α−H and δ−H protons. Thus, the proline coordination is chemioselective in contrast to the molecules, shown in Table 12.4. Here, on addition of complex 1 the relaxation rates of all the protons increase similarly.

Table 12.2　$^1H T_1$ times (s) in proline molecules measured at 200 MHz in the absence and the presence of Cu(II) complex 1 in aqueous solution. (Reproduced with permission from Y. N. Belokon' et al. *Journal of the Chemical Society Dalton Transactions* 1990; 1873. © The Royal Society of Chemistry)

Time (295 K)	N_{Cu}/N_P	α	β1	γ + β2	δ2	δ1
T_2	0	2.53	2.20	2.15	1.93	1.84
T_2	5×10^{-4}	0.18	0.47	0.57	0.22	0.21
T_1	0	3.44	1.91	2.77	2.22	2.02
T_1	10^{-4}	0.92	1.66	1.96	1.37	0.93
T_1	5×10^{-4}	0.20	0.88	1.07	0.48	0.26
T_1	10^{-3}	0.11/0.12	0.65/0.63	0.79/0.77	0.29/0.31	0.16/0.16
T_1	5×10^{-3}	0.08/0.08	0.54/0.47	0.68/0.57	0.22/0.21	0.12/0.10

Table 12.3　Changes in the ^1H spin–lattice relaxation rates in proline caused by addition of Cu(II) complex 1, calculated using the values from Table 12.2. (Reproduced with permission from Y. N. Belokon' et al. *Journal of the Chemical Society Dalton Transactions* 1990; 1873. © The Royal Society of Chemistry)

N_{Cu}/N_P	α	β1	γ + β2	δ2	δ1
10^{-4}	0.80	0.08	0.15	0.28	0.58
5×10^{-4}	4.64	0.61	0.58	1.63	2.35
10^{-3}	8.48/8.16	1.03/1.08	0.91/0.94	3.02/2.83	5.92/5.86
5×10^{-3}	11.76/12.53	1.34/1.63	1.12/1.39	4.12/4.36	8.20/9.41

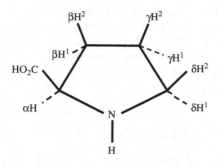

Figure 12.5 The structure of the proline molecule (schematically)

Table 12.4 Spin–lattice relaxation times (s) for the protons of different compounds in the presence of Cu(II) complex 1 at $N_{Cu}/N_P = 5 \times 10^{-4}$ (Reproduced with permission from Y. N. Belokon' *et al. Journal of the Chemical Society Dalton Transactions* 1990; 1873. © The Royal Society of Chemistry)

Compound	α	β	γ
Prolylamine	5.51	5.60	5.43
Prolylamine/1	0.62	1.15	2.03
Propanoic acid	5.00	6.17	
Propanoic acid/1	3.23	3.77	
Butanoic acid	2.61	2.73	2.82
Butanoic acid/1	2.04	2.16	2.27

By assuming fast exchange (Equation 12.10), Equation (12.13) is rewritten as:

$$1/T_{1M} = (N_P/N_{Cu}Q)(1/T_1 - 1/T_{1F}) \qquad (12.14)$$

where T_1 is the measured relaxation time; T_{1M} and T_{1F} are the relaxation times in the paramagnetic proline/complex 1 and in the free proline molecule, respectively; N_P and N_{Cu} are the molar concentrations of the proline and complex 1; Q is the proline/complex 1 ratio. The T_{1M} values, calculated from the data of Table 12.3, are constant in the range $N_{Cu}/N_P = 10^{-4}-10^{-3}$. This result supports the assumption of a fast exchange of the proline in the coordination sphere of complex 1. When the N_{Cu}/N_P ratio is larger than $5 \times 10^{-3} : 1$, the T_{1M} time begins to change, indicating the appearance of uncoordinated Cu(II) centers in the solution.

Since the electron spin relaxation time for the copper ion is 10^{-8} s and the magnitude of $(A/\hbar)^2$ can be taken as $10^{12}-10^{14}$ Hz, and since the lifetime of a nucleus in the first coordination sphere of the metal may safely be assumed to be greater than 10^{-8} s, then contact contribution $R_{1,2}^{CONT}$ is calculated as

10^1 s^{-1}. The estimated magnitude and the insignificant differences between the T_1 and T_2 times in Table 12.2 show the contact contributions to be negligible. Thus, the T_{1M} relaxation times, obtained via Equation (12.14), can be used for calculations of interatomic distances.

The proton–copper and carbon–copper distances are calculated via the equation:

$$r(\text{Å}) = C \{T_{1M}[3\tau_C + 7\tau_C/(1 + (1.71 \times 10^{19}\, \omega_I^2\, \tau_C^2))]\}$$

$$1/\tau_C = 1/\tau_E + 1/\tau_M + 1/\tau_R \qquad (12.15)$$

as a simplified form of Equation (12.4), where $C = 539$ and 340 for protons and carbons, respectively. It has been mentioned that the τ_C values for small and medium-size ligands, coordinated with Cu (II) centers, are controlled by the rotational molecular correlation times τ_R. For systems with molecular sizes similar to the 1:1 proline/complex 1 adduct, the τ_R value is 10^{-10} s. Then, the T_{1M} times lead to r(H–Cu) distances shown in Table 12.5. As can be seen, these distances are independent of the concentration of complex 1 up to $N_{Cu}/N_P = 5 \times 10^{-3}$. Finally, the structure of the adduct, created on the basis of the H–Cu distances (Figure 12.6), agrees well with the C–Cu distances

Table 12.5 Distances (Å) between the proline protons and Cu(II) of complex 1 in the mixed complex. (Reproduced with permission from Y. N. Belokon' *et al. Journal of the Chemical Society Dalton Transactions* 1990; 1873. © The Royal Society of Chemistry)

N_{Cu}/N_P	α	$\beta 1$	$\gamma + \beta 2$	$\delta 2$	$\delta 1$
10^{-4}	3.38 ± 0.25	4.96 ± 0.35	4.46 ± 0.35	4.02 ± 0.3	3.56 ± 0.25
5×10^{-4}	3.29 ± 0.25	4.62 ± 0.35	4.66 ± 0.35	3.92 ± 0.3	3.49 ± 0.25
10^{-3}	3.37 ± 0.25	4.71 ± 0.35	4.82 ± 0.35	4.01 ± 0.3	3.56 ± 0.25
5×10^{-3}	4.10 ± 0.25	5.76 ± 0.35	5.91 ± 0.45	4.89 ± 0.4	4.30 ± 0.35

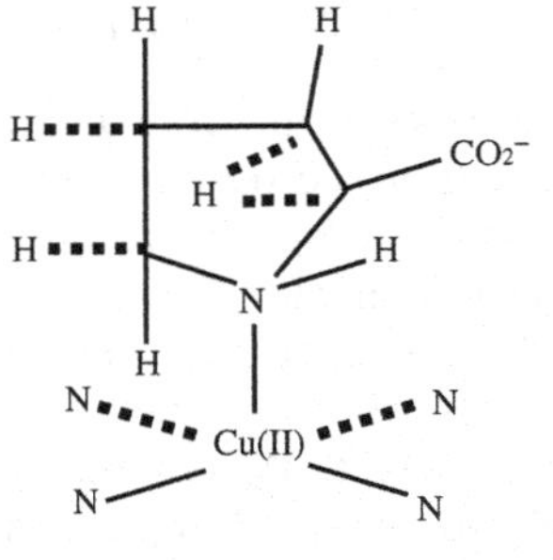

Figure 12.6 The fragment of the complex formed by proline and ({(R)-7,8,15,18-tetrahydro-5,10-bis(2'-methoxyethoxy-5'-methylphenyl)-7-(p-trimethylammoniophenyl)dibenzo[e, m]-[1,4,8,11]tetra-azacyclotetradecine-16,17-dionato(1-)-N^{6}N^{9}N^{15}N^{18}}-copper(II) chloride. The structure is established by the ^{1}H and ^{13}C NMR relaxation measurements in D$_2$O

Table 12.6 ^{13}C spin–lattice relaxation times (s) of proline in the absence (T_{1F}) and the presence of Cu(II) complex 1 (T_1) and C–Cu distances r (Å) in the mixed complex at $N_{Cu}/N_P = 5 \times 10^{-4}$

	α	β	γ	δ	CO_2^-
T_{1F}	6.43	5.02	6.81	5.29	100
T_1	2.55	3.90	4.68	2.55	15
r	3.1	4.0	3.7	3.2	4.0

evaluated from the ^{13}C T_1 times (Table 12.6). Note that the structure exhibits a preference for an envelope conformation of proline molecules.

12.4 Kinetics of Ligand Exchange via Paramagnetic Relaxation Rate Enhancement

As we have shown, chemical exchange processes affect the relaxation rate enhancements. If $r(I-S)$ distances, electron spin relaxation times and the constants A are known, then relaxation measurements directly result in exchange rates. However, under some conditions and assumptions, the kinetic parameters of exchange processes can be obtained, even in the absence of $r(I-S)$, τ_E and A data. This approach is based on T_1 and T_2 times, determined at different temperatures and different concentrations of paramagnetic ions, and on measurements of *paramagnetic shifts* in the NMR spectra. Hyperfine interactions with the unpaired electrons affect chemical shifts of neighboring nuclei and this influence is measured as paramagnetic shifts. Usually, the paramagnetic shifts consist of the Fermi contact (FC) contributions caused by appearance (delocalization) of the unpaired spin density on target nuclei, and the pseudo-contact shifts, (PC) due to through-space electron–nucleus interactions:

$$\delta = \delta_{FC} + \delta_{PC} \qquad (12.16)$$

^{13}C NMR experiments in aqueous solutions of D-gluconic acid (compound L) and the copper(II) perchlorate (compound M) forming the complex shown in Figure 12.7 [6], illustrate the method of investigation. The variable-temperature ^{13}C T_1, T_2 times and ^{13}C paramagnetic shifts ($\Delta\omega$, expressed in frequency units) have been obtained at different concentrations C_L where $C_L \gg C_M$. Since in the presence of a ligand exchange all the measured magnitudes depend on the sets of parameters at both paramagnetic (T_{1M}, T_{2M} and $\Delta\omega_M$) and diamagnetic (T_{1F}, T_{2F} and 0) sites and on exchange rate $1/\tau_M$,

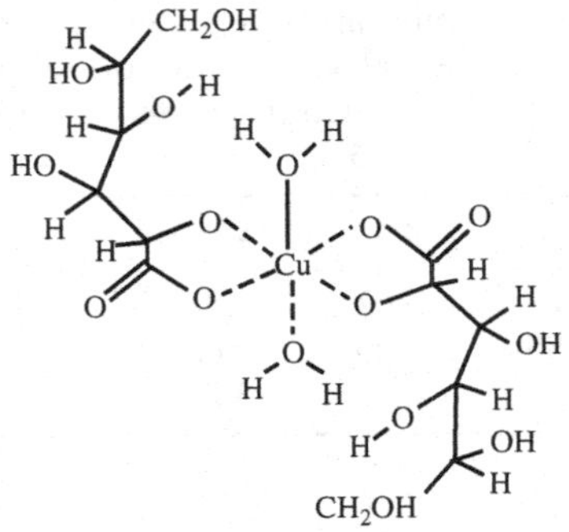

Figure 12.7 Structure of the complex formed by D-gluconic acid with Cu(II) in a water solution. (Reproduced from T. Gajda *et al. Inorganica Chimica Acta* 1998; **275**, 130, with permission from Elsevier)

following Swift and Connick one can write:

$$1/T_{1r} = (1/T_1 - 1/T_{1F})/P_M = 1/(T_{1M} + \tau_M)$$

$$1/T_{2r} = (1/T_2 - 1/T_{2F})/P_M = (1/\tau_M)\{(1/T_{2M}^2 + 1/T_{2M}\tau_M + \Delta\omega_M^2)/$$

$$((1/T_{2M} + 1/\tau_M)^2 + \Delta\omega_M^2)\} \qquad (12.17)$$

Here $P_M = 2C_M/C_L$ and $\Delta\omega_M$ is the chemical shift in the paramagnetic complex, formed by L and Cu(ClO$_4$)$_2$. If the pseudo-contact contribution is negligible, then the paramagnetic chemical shift can be expressed via the equations:

$$\Delta\omega_r = \Delta\omega/P_M = \Delta\omega_M \{(1 + \tau_M/T_{2M})^2 + \Delta\omega_M^2\tau_M^2\}^{-1} \qquad (12.18)$$

$$\Delta\omega_M/\omega_C = 2\pi A_C \mu_M (S(S+1))^{-1/2}/3k\gamma_C T \qquad (12.19)$$

where μ_M is the electronic magnetic moment and ω_C is the Larmor frequency of ^{13}C nuclei. Note that the magnitudes of $1/T_{1r}$, $1/T_{2r}$ and $\Delta\omega_r$ can be determined experimentally from least-squares plots of $1/T_1$, $1/T_2$ and $\Delta\omega$ versus P_M at constant temperature.

Rates of nuclear relaxation, $1/T_{1M}$ and $1/T_{2M}$, in the paramagnetic complex can be expressed via Equations (12.4). The equations are greatly simplified at $\tau_M \gg \tau_E \gg \tau_R$ and $\omega_E^2\tau_E^2 \ll 1$ converting to:

$$1/T_{1M} = 2 \times 10^{-14} \mu_M^2\gamma_C^2 \tau_R/5r_{(C-M)}^6 (1 + \omega_C^2\tau_R^2)$$

$$1/T_{2M} = (7/6) (1/T_{1M}) + 4\pi^2 A_C^2 S(S+1) \tau_E/3 \qquad (12.20)$$

Equations (12.20), (12.17) and (12.19) can be combined with $\tau = \tau_0 \exp(\Delta E/RT)$ (for τ_M and τ_R times) to express the dependences of the T_{1r}, T_{2r} and $\Delta\omega_r$ parameters on the temperature. Note that the electron relaxation rate $1/\tau_E$

Table 12.7 ^{13}C T_1 relaxation date, carbon–copper distances and A constants obtained for complex CuL_2 of D-gluconic acid in water at 298 K

Parameter	C(1)	C(2)	C(3)	C(4)	C(5)	C(6)
$1/T_{1r}$ (s^{-1})	2175	2410	502.5	158	132	52.5
$1/T_{1M}$ (s^{-1})	2987	3449	536	161	134	53
r(C–M) (Å)	2.81	2.76	3.72	4.52	4.96	5.79
A (MHz)	0.98	0.97	0.17	0.13		

can be taken in these expressions as K/τ_R with the constant K independent of the temperature and the frequency. It is obvious, however, that direct fitting of the experimental data to such theoretical equations is impossible because of a great number of unknown parameters: A_C, τ_{0M} and τ_{0R}, ΔE_R, ΔE_M, K and r(C–M). However, a search of the parameters can be narrowed by increasing a number of nuclei involved in the relaxation experiments.

An analysis of the ^{13}C NMR data, collected for L at variations in C_M concentrations, shows a very important feature: the ligand exchange is *slow* on the NMR time scale for the C(1) and C(2) carbons ($\Delta\omega_M^2\tau_M^2 \gg 1$) and moderately *fast* for the C(3) and C(4) resonances ($\Delta\omega_M^2\tau_M^2 \sim 1$). In the first case $1/T_{2r} = 1/\tau_M$ and thus the τ_M times can be calculated from the spin–lattice relaxation measurements at any temperature. Then, the standard temperature set of the τ_M values gives τ_{0M} (3.0×10^{-12} s) and activation energy parameters of the exchange: $\Delta E_M = 43.4$ kJ/mol, $\Delta H^{\neq} = 45.9$ kJ/mol, $\Delta S^{\neq} = -7.6$ e.u.

The paramagnetic shifts, $\Delta\omega_M$, of C(3) and C(4) nuclei can be obtained simply as a high-temperature limit for Equation (12.18) (i.e. $\Delta\omega_r = \Delta\omega_M$) and thus the corresponding A_C values can be calculated. Then, the $1/T_{1M}$ values are determined as $1/T_{1M} = (T_{1r} - \tau_M)^{-1}$ and thus all the data lead to C–M distances, listed in Table 12.7. The structure of the paramagnetic complex, deduced on the basis of the C–M distances, is shown in Figure 12.7. Finally, a similar approach can be applied to more complex molecules, for example, Gd(III)- and Mn(II)-coordinated adenosine triphosphate, studied by ^1H and ^{31}P relaxation measurements [7] or Cu(II)-coordinated histidine-containing peptides [1].

12.5 Longitudinal Electron Relaxation Time at Paramagnetic Centers from Variable-high-field NMR Experiments

The theory emphasizes that relaxation of electrons, located at paramagnetic centers, plays an important role in the NMR behavior of nuclei neighboring these centers. Moreover, under some conditions, quantitative interpretation of T_1 and T_2 times, measured in the presence of paramagnetic species, is impossible without precise knowledge of the electron relaxation times. On the other hand, knowledge of electronic relaxation can give chemically significant information, particularly in the case of biological systems. It is known, for example, that the biological function of metalloproteins is governed by the geometric and electronic features of their metal sites. In turn, these features affect relaxation of the unpaired electrons. Thus, measuring the electron relaxation rates in metalloproteins is a direct way of understanding the structure and biological function of such systems.

Nuclear relaxation in Cu(II)-containing protein molecules can be expressed in general form [8] as:

$$R_{1\rho} = (2/5)(\mu_0/4\pi)^2\gamma_I^2 g_e^2\beta_e^2 S(S+1)\Delta^2(3\tau_C/(1+\omega_I^2\tau_C^2)) \tag{12.21}$$

where τ_C has its usual meaning ($\tau_C^{-1} = 1/\tau_{1E} + 1/\tau_R + 1/\tau_M$) and Δ is a parameter, depending on metal–nucleus distances and the fraction of the unpaired electron spin delocalized to the ligand. The single field-dependent component in this equation is $1/\tau_{1E}$. For this reason, if rates of nuclear relaxation change in variable-field NMR experiments, then electron relaxation times can be directly determined from the $R_{1\rho}$ values.

Ma and Led [8] demonstrate the effectiveness of this approach with the help of ^1H, ^{13}C NMR experiments on blue copper proteins. Figure 12.8

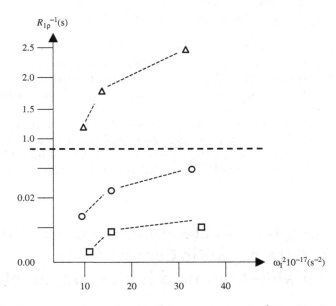

Figure 12.8 Frequency dependence of the paramagnetic ^1H relaxation rates for: Leu(14) H^δ (squares); Val(41) H^γ (circles); and Leu(59) H^δ (triangles) in *Anabaena variabilis plastocyanin* (schematic). Data for 400, 500 and 750 MHz

displays the field effects obtained for some protons in *Anabaena variabilis plastocyanin*. As can be seen, the plots of the $R_{1\rho}$ versus ω_I^2 are not linear. Furthermore, similar effects are observed for ^{13}C nuclei. This nonlinearity is compatible only with a variation of τ_C at the given field diapason. According to independent data, the exchange rate, $1/\tau_M$, is small in the protein molecule and therefore it can be ignored. The molecular rotation correlation time τ_R is as long as 6.2 ns at 298 K. Thus, the electron relaxation rate dominates in the τ_C^{-1} term. It is easy to show that the ratio $R_{1\rho}$ (11.7 T)$/R_{1\rho}$ (17.6 T) for a target nucleus depends only on the value of the spectral density function. Thus, only two different $R_{1\rho}$ (11.7 T)$/R_{1\rho}$ (17.6 T) values are needed for the $1/\tau_{1E}$ determination: one from a ^{13}C nucleus and one from ^1H nucleus. Then, the electron spin relaxation rates at the two magnetic fields can be calculated by a least-squares fit of Equation (12–21) to the experimental $R_{1\rho}$ (11.7 T)$/R_{1\rho}$ (17.6 T) ratios obtained for ^{13}C and ^1H nuclei. Finally, the calculations give $1/\tau_{1E}$ values of 5.8×10^9 and 2.6×10^9 s^{-1} at 11.7 and 17.6 T, respectively. This method does not require knowledge of the structure of the proteins.

Bibliography

1. E. Gaggelli, N. D'Amelio, D. Valensin, G. Valensin, *Magnetic Resonance in Chemistry* 2003; **41**: 877.

2. I. Bertini, C. Luchinate. NMR of paramagnetic substances. *Coordination Chemistry Reviews*, 1996; **150**: 1.

3. R. Sharp, L. Lohr, J. Miller. *Progress in Nuclear Magnetic Resonance Spectroscopy* 2001; **38**: 115.

4. L. Banci, C. Luchinat. *Inorgica Chimica Acta* 1998; **275**: 373.

5. Y. N. Belokon, L. K. Pritula, V. I. Bakhmutov, D. G. Gusev, V. M. Belokov. *Journal of the Chemical Society Dalton Transactions* 1990; 1873.

6. T. Gajda, B. Gyurcsik, T. Jakusch, K. Burger, B. Henry, J.-J. Delpuech. *Inorgica Chimica Acta* 1998; **275**: 130.

7. L. V. Elst, R. N. Muler. *Inorgica Chimica Acta* 1998; **272**: 92.

8. L. Ma, J. J. Led. *Journal of the American Chemical Society* 2000; **122**: 7823.

Concluding Remarks

Because of the wide range of applications of nuclear magnetic resonance in chemical practice, many undergraduate chemistry courses include the theoretical basis of the NMR spectroscopy. However, a chemistry course is not complete if it does not consider and discuss the basic principles of NMR relaxation. This book attempts to explain the relaxation phenomenon from the point of view of a chemist, and focuses on the theory and methodological aspects of nuclear relaxation as an independent and powerful tool in chemical studies. Discussing relaxation approaches to different chemical topics and interpretations of relaxation data, we have tried to avoid the situations when graduate students and postdoctoral fellows consider nuclear relaxation as a 'black box.' Thus, this material can be easy incorporated into undergraduate and graduate teaching.

Finally, for more advanced reading we recommend the following original reviews and papers:

Bibliography

1. P. Luginbuhl, K. Wuthrich. Semi-classical nuclear spin relaxation theory revisited for use with biological macromolecules. *Progress in Nuclear Magnetic Resonance Spectroscopy* 2002; **40**: 199.

2. A. V. Zakharov, R. Y. Dong. Rotational viscosity, dynamic phenomena and dielectric properties in a long-chain liquid crystals: NMR study and theoretical treatment. *Physical Review E* 2000; **63**(1): 1704.

3. F. C. Marincola, M. Casu, G. Saba, A. Lai. ^{23}Na NMR relaxation studies of the Na-DNA/drug interactions. *ChemPhysChem. A European Journal of Chemical Physics and Physical Chemistry* 2001; **2**: 569.

Practical NMR Relaxation for Chemists Vladimir I. Bakhmutov
© 2004 John Wiley & Sons, Ltd ISBNs: 0-470-09445-1 (HB); 0-470-09446-X (PB)

4. A. J. Horsewill. Quantum tunneling aspects of methyl group rotation studies by NMR. *Progress in Nuclear Magnetic Resonance Spectroscopy* 1999; **35**: 359.

5. F. A. A. Mudler, M. Akke. Carbonyl ^{13}C transverse relaxation measurements to sample protein backbone dynamics. *Magnetic Resonance Chemistry* 2003; **41**: 853.

6. N. Murali, V. V. Krishnan. A primer for nuclear magnetic relaxation in liquids. *Concepts in Magnetic Resonance* 2003; **17A**: 86.

7. D. Frueh. Internal motions in proteins and interference effects in nuclear magnetic resonance. *Progress in Nuclear Magnetic Resonance Spectroscopy* 2002; **41**: 305.

8. A. Bagno, G. Scorrano. Selectivity in proton transfer, hydrogen bonding, and solvation. *Accounts of Chemical Research* 2000; **33**: 609.

9. P. B. Kingsley. Signal intensities and T_1 calculations in multiple-echo sequences with imperfect pulses. *Concepts in Magnetic Resonance* 1999; **11**: 29.

10. D. A. Case. Molecular Dynamics and NMR spin relaxation in proteins. *Accounts of Chemical Research* 2002; **35**: 325.

11. A. Ross, M. Czisch, G. C. King. Systematic errors associated with the CPMG pulse sequence and their effect on motional analysis of biomolecules. *Journal of Magnetic Resonance* 1997; **124**: 355.

12. J. Boyd. Measurement of ^{15}N relaxation data from the side chains of asparagines and glutamine residues in proteins. *Journal of Magnetic Resonance Series B* 1995; **107**: 279.

13. D. M. Lemaster, D. M. Kushlan. Dynamical mupping of *E. coli* thioredoxin via ^{13}C NMR relaxation analysis. *Journal of the American Chemical Society* 1996; **118**: 9263.

14. N. A. Farrow, O. Zhang, A. Szabo, D. A. Torchia, L. E. Kay. Spectral density function mapping using ^{15}N relaxation data exclusively. *Journal of Biomolecular NMR* 1995; **6**: 153.

15. B. Brutsscher. Principles and applications of cross-correlated relaxation in biomolecules. *Concepts in Magnetic Resonance* 2000; **12**: 207.

16. A. Kumar, R. C. R. Grace, P. K. Madhu. Cross-correlations in NMR. *Progress in Nuclear Magnetic Resonance Spectroscopy* 2002; **37**: 191.

17. H. Mo, T. C. Pochapsky. Intermolecular interactions characterized by nuclear Overhauser effects. *Progress in Nuclear Magnetic Resonance Spectroscopy* 1997; **30**: 1.

18. O. Soubias, V. Reat, O. Saurel, A. Milon. ^{15}N T_2 relaxation times of bacteriorhodopsin transmembrane amide nitrogens. *Magnetic Resonance Chemistry* 2004; **42**: 212.

19. I. Bertini, C. Luchinat, G. Parigi (eds). Solution NMR of paramagnetic molecules. *Current Methods in Inorganic Chemistry* 2001; **2**: 1–376.

Index

Practical NMR Relaxation for Chemists   Vladimir I. Bakhmutov
© 2004 John Wiley & Sons, Ltd   ISBNs: 0-470-09445-1 (HB); 0-470-09446-X (PB)